CONFRONTING THE CLIMATE CHALLENGE

LAWRENCE H. GOULDER
AND
MARC A. C. HAFSTEAD

CONFRONTING THE CLIMATE CHALLENGE

U.S. Policy Options

COLUMBIA UNIVERSITY PRESS
NEW YORK

Columbia University Press
Publishers Since 1893
New York   Chichester, West Sussex
cup.columbia.edu

Library of Congress Cataloging-in-Publication Data
Names: Goulder, Lawrence H. (Lawrence Herbert), author. |
Hafstead, Marc A. C., author.
Title: Confronting the climate challenge : U.S. policy options
/ Lawrence H. Goulder and Marc A. C. Hafstead.
Description: New York : Columbia University Press, [2018] | Includes bibliographical
references and index.
Identifiers: LCCN 2017030796 (print) | LCCN 2017033236 (ebook) |
ISBN 9780231545938 | ISBN 9780231179027 (alk. paper)
Subjects: LCSH: Climatic changes—Government policy—United States. |
Emissions trading—Government policy—United States. | Greenhouse gas
mitigation—Government policy—United States.
Classification: LCC QC903.2.U6 (ebook) | LCC QC903.2.U6 G68 2018 (print) |
DDC 363.738/745610973—dc23
LC record available at https://lccn.loc.gov/2017030796

Columbia University Press books are printed on permanent
and durable acid-free paper.
Printed in the United States of America

Cover design: Chang Jae Lee
Cover image: Richard Newstead © GettyImages

To Our Parents

CONTENTS

III: POLICY APPROACHES AND OUTCOMES

IV: CONCLUSIONS

PREFACE

This might seem to be an inauspicious time to offer a book on national policy options for confronting climate change. At the time of this writing, there is considerable political opposition at the U.S. federal level to climate-change policies.

Our view is that the seriousness of the climate-change problem justifies the effort, despite the political challenges. Climatologists and other scientists have reached a strong consensus that human activities are contributing significantly to climate change, and that in the absence of significant efforts to reduce future emissions of greenhouse gases, future changes in climate will be substantial and the associated damages severe. Under these circumstances, it seems more important than ever to assess what confronting the issue implies for the U.S. economy.

Two sources of opposition to climate policy—beyond the disbelief that human activities are driving climate change—are the convictions that efforts to address the problem will cost businesses and households too much relative to the benefits, and that the economic sacrifices stemming from these efforts will be unfairly distributed. This book addresses these sources of opposition. Our analyses point the way to a range of policies whose benefits—avoided climate damages and associated beneficial health impacts—substantially exceed the costs. Moreover, these analyses indicate that when carefully designed, several policies avoid placing unfair burdens on low-income groups or causing profit losses to particularly vulnerable industries. The impacts of climate policies cannot be predicted with precision: there are important uncertainties. But these general findings emerge under a range of alternative assumptions that address the uncertainties.

Although the prospect of serious climate change is likely to be with us for a long time, focusing on the issue in the near term can help prevent substantial future harm, given the time lags and irreversibilities in the climate system. We hope that the information in this book will elevate the debate about U.S. climate policy and catalyze productive efforts at the federal level.

ACKNOWLEDGMENTS

Over the years, we have benefited tremendously from the outstanding advice and encouragement we have received from many people. Well before the work on this book began, Stanford colleagues John Shoven, Jim Sweeney, and the late Alan Manne served as inspiring and friendly mentors, providing the key intellectual foundations for the general-equilibrium (multimarket) framework applied here. Ray Kopp, George Shultz, and John Weyant are longtime close followers of our work who continue to provide very helpful ideas. We are very grateful to these colleagues for their insights and support.

We are deeply indebted to Lans Bovenberg, Dallas Burtraw, Yunguang Chen, Mark Jacobsen, Koshy Mathai, Ian Parry, the late Steve Schneider, Rob Stavins, Arthur van Benthem and Rob Williams, with whom we have collaborated on U.S. climate policy research. It has been a great pleasure to work with these colleagues. Their input has provided the basis for much of what is contained in the book and has significantly raised the quality of our work. We are also grateful to Pete Wilcoxen for guiding us through some of the important econometric challenges we have faced.

We are happy to thank Chris Bruegge, Derek Gurney, Gyurim Kim, Xianling Long, Emelia von Saltza, Ishuwar Seetharam, and Santiago Saavedra for their excellent research assistance along many dimensions, including data development, model construction, model application, and interpretation of model results. We thank Bridget Flannery-McCoy, our main contact at Columbia University Press, for consistently helpful advice, thoughtful and flexible responses to our questions and suggestions, and very close attention to our manuscript throughout the publication process. Thanks also to Rebecca Edwards of Cenveo Publisher Services for sustained help with the formatting of text, tables, and graphs. We are grateful to the Energy Foundation, to Stanford University's Hoover Energy Task Force and its Natural Gas Initiative, and to Resources for the Future's Carbon Tax Initiative for financial support.

Finally, we thank our partners Martha Shirk and Ashley Hafstead for moral support, editorial advice, and tolerance during the times when the book took too much of our attention.

CONFRONTING THE CLIMATE CHALLENGE

U.S. Policy Options

I

INTRODUCTION AND ANALYTICAL BACKGROUND

1

INTRODUCTION

I n recent decades, the average surface temperature of the Earth has continued to set record highs. Natural scientists have mounted increasing
evidence that observed increases in temperature are due in significant
part to the increase in atmospheric concentrations of greenhouse gases
resulting from human activities, including the combustion of carbon-based
fuels. The most recent assessment report of the UN-sponsored Intergovernmental Panel on Climate Change concluded that it was "extremely likely"
(i.e., having a probability of 95–100 percent) that significant changes in
climate have been and would continue to be driven by human activity.[1]

There is considerable uncertainty about both the trajectory of future
emissions and the extent to which Earth's climate will continue to change
as a consequence. However, there is a clear consensus among climate scientists that in the absence of significant reductions of greenhouse gas
emissions from anticipated business-as-usual levels, future climate change
will be extensive and produce major damage to the economy and the environment. A central estimate is that, under business as usual, the sea level
will rise by 2 to 3 feet by the end of this century and continue to rise for
centuries after, flooding low-lying coastal regions and causing displace-
ments of human populations throughout the world.[2] It is also predicted
that the tropics will expand, leading to greater prevalence of tropical dis-
eases.[3] Under business as usual, snowpacks are expected to shrink or disappear, which would substantially raise the costs of supplying water for
crop irrigation or urban use during summer months.

The threat of such climate-related damages warrants a close look at
policy options to reduce or avoid these damages. Policies differ in the

extent to which they would reduce emissions of greenhouse gases, in their overall costs to the economy, and in the ways the costs are distributed across different industries or different groups of people (e.g., households with different levels of income). Sensible climate-change policy requires a careful analysis of the anticipated impacts along these and other dimensions.

Given the significant time lags in the climate system, as well as the long atmospheric lifetimes of greenhouse gases (and associated irreversibilities), the extent of future climate damage is determined by decisions societies make now. This makes it important to have good information today to guide policy choices. Delay poses risks.

This book explores the strengths and weaknesses of a range of federal policy options for reducing emissions of carbon dioxide (CO_2), the most important greenhouse gas. Although tackling the problem of climate change surely requires efforts by countries around the globe, the U.S. can play a major role in addressing the problem.

Current climate-change policy in the U.S. includes efforts at the municipal and state levels. More than 1,000 mayors recently reaffirmed their commitment to the U.S. Conference of Mayors' Climate Protection Agreement, according to which cities have pledged to reduce their carbon emissions below 1990 levels. Some of the most ambitious efforts are being undertaken by the states. California regulates CO_2 emissions through a suite of policies that includes a cap-and-trade program, and nine states in the Northeast participate in the Regional Greenhouse Gas Initiative (RGGI), a cap-and-trade system regulating power plant CO_2 emissions. In addition, thirty states have implemented renewable portfolio standards (RPSs) that relate to purchases of wholesale electricity by electric utilities. These standards help reduce CO_2 emissions by imposing a floor on the fraction of utilities' purchases of electricity coming from renewable sources such as solar panels and wind farms.

At the federal level, there has been much debate for decades on climate policy but as yet no significant legislative action. In 2009, the House of Representatives passed the American Clean Energy and Security Act (also known as the Waxman-Markey bill), which would have created a national cap-and-trade program for greenhouse gases. But the Senate never voted on this proposal and the U.S. Congress has taken no action on cap and trade since. In his 2011 State of the Union address, President

Obama expressed support for a Clean Energy Standard (CES) to increase the share of electricity coming from clean sources, and in 2012 former Senator Jeff Bingaman (D-NM) proposed a bill involving a CES, but the proposal failed to gain traction in Congress.

In the absence of congressional action on climate-change policy, the Obama administration launched efforts to address climate change through the executive branch. That administration proposed increasingly stringent standards on automobile fuel economy, raising the Corporate Average Fuel Economy (CAFE) standards for new light-duty cars and trucks to 54 miles per gallon by 2025, a significant increase from the 2016 average of about 26 miles per gallon. Although the CAFE standards pertain to gasoline use, they help reduce emissions of CO_2 since reduced gasoline combustion implies lower CO_2 emissions. In addition, in 2014 the Obama administration proposed the Clean Power Plan (CPP), an effort to control emissions of CO_2 from existing fossil-based electricity generators under the authority of the Clean Air Act.

With the election of President Trump, executive branch support for climate-change policy vanished. In March 2017, President Trump instructed the EPA "to as soon as practicable, suspend, revise, or rescind" the CPP. The new administration also appears to oppose the tightening of CAFE standards sought by the Obama administration. And in June 2017, President Trump announced that the U.S. would pull out of the December 2015 Paris Agreement, in which 195 nations had pledged to reduce greenhouse gas emissions.[4]

With these statements and actions by the Trump administration, federal support for climate policy seemed to reach a new low. Nevertheless, there remains significant interest and movement on climate policy in many U.S. states. Indeed, shortly after President Trump announced the plans for the U.S. pull-out, numerous U.S. cities and states pledged further action on climate. For example, governors Jerry Brown of California, Jay Inslee of Washington, and Andrew Cuomo of New York announced a new coalition of states pledging to uphold the Paris targets. And in June 2017, 323 U.S. mayors, representing 62 million Americans, signed a pledge to intensify their cities' current climate goals and "adopt, honor and uphold" the commitments of the Paris agreement. Support for climate policy is not limited to Democratic governors and mayors, either. In February 2017, a group of eight very distinguished Republican statesmen, under the umbrella of the Climate Leadership Council (CLC), promoted "The Conservative Case for Carbon Dividends," a policy brief that pushed for legislation that would introduce a carbon tax as a replacement for Obama-era regulations. In June 2017, the CLC announced its founding members, including international oil firms ExxonMobil, BP, Shell, and Total along with distinguished individuals such as Michael Bloomberg and Stephen Hawking.

Moreover, much of the American public continues to support U.S. climate policy. In a January 2015 poll conducted by Resources for the Future, *The New York Times*, and Stanford University, 78 percent of respondents agreed that the "government should limit the amount of greenhouse gases that the U.S. businesses put out." A February 2017 poll by the Yale Program on Climate Change Communication indicates that 82 percent of Americans favor regulating CO_2 as a pollutant. According to Gallup surveys, the percentage of American adults that believe global warming is caused by human activities has risen from 55 percent to 68 percent from 2015 to 2017, and the percentage that "worry a great deal" about global warming has increased from 32 percent to 45 percent over the same interval.

Concerns about climate change, and support for climate-change policy, are growing. The basic predictions that climate scientists have made over the past two decades about the augmentation of climate change have been borne out so far. If the scientists' predictions continue to be correct— that is, if the changes to the climate become increasingly severe and the globe experiences worsening impacts of such change—political support for federal-level action to confront this phenomenon could increase significantly. In this situation, carefully derived information on the potential economic impacts of alternative climate-change policies can be of great value. Such information can elevate policy discussions and help catalyze action at the national level.

OUR FOCUS

This book aims to provide valuable information about the impacts of important U.S. national policy options. It considers a range of options, concentrating on policies that are broad and consistent across sectors and regions of the economy. Breadth and consistency are key to delivering significant emissions reductions at low cost. Broad coverage helps reduce incentives by businesses or households to undertake costly efforts to relocate in order to escape the stringent state or local regulations.[5] And consistency helps avoid significant differences in the marginal costs of achieving emissions reductions across firms or regions.[6] As discussed in later chapters, such differences usually indicate that emissions reductions are not being carried out where they can be accomplished at the lowest cost.

In the following chapters, we examine both the overall cost of these policies and how those costs are distributed across industries, households, and regions of the country. The distribution of policy impacts is highly relevant to the fairness and political feasibility of the various options.

The federal options explored span a range of policy approaches. One is a carbon tax—a tax on fossil fuels in proportion to the carbon content. In nearly all uses, the carbon content of the fossil fuel determines the amount of CO_2 that will be emitted when the fuel or its derivative refined fuel is combusted.[7] Thus, a carbon tax is a form of emissions pricing: it is implicitly a tax on the emissions stemming from fuel combustion.

A second option examined is a cap-and-trade program. Under cap and trade, the government issues a given number of emissions allowances (or permits) to various firms or facilities, with each allowance entitling the owner to a given quantity of emissions over some interval of time. Allowances can be traded. The sales (supply) and purchases (demand) of allowances yield a price of allowances in the market. This implies that firms face a price on emissions at the margin (i.e., for the last unit of emissions generated), since every additional unit of emissions by a given firm obligates that firm either to purchase an additional allowance (if it doesn't already have enough allowances to justify its intended emissions) or to sell one less allowance (if it currently has more allowances than it needs to justify its planned emissions).

A third policy is a federal CES. As mentioned earlier, the CES is a requirement relating to utilities' purchases of electricity on the wholesale market. This policy contrasts with the other policies we consider in that it is not a form of emissions pricing: it imposes a constraint on utilities' purchases rather than a price on CO_2. Under a CES, the share represented by "green" sources (e.g., wind-powered generation and solar-powered generation) must not fall below some specified floor. This form of regulation would represent at the national level a policy similar in structure to the renewable portfolio standards already in place in a large number of states.[8] As mentioned earlier, a CES policy was proposed as recently as 2012.

The fourth policy is an increase of the federal tax on gasoline. While legitimately categorized as a form of emissions pricing, a gasoline tax introduces a price on only the emissions associated with gasoline combustion. It does not cover the greenhouse gas emissions resulting from other refined products of fossil fuels, such as kerosene, benzene, and jet fuel. Nor does it cover emissions from the combustion of coal or natural gas. Hence it is narrower in coverage than a typical carbon tax or cap-and-trade system. Some policy analysts regard the relatively narrow base as a drawback, in line with the idea that the broader the base of a tax on emissions, the greater its potential to pick the "low-hanging fruit" (i.e., to exploit the low-cost options for reducing emissions). However, a potential attraction of a gasoline tax is its ability to address several "local" environmental problems that are intimately connected with gasoline use—problems including traffic congestion and health effects from other pollutants associated with gasoline combustion. In our assessment of an increased gasoline tax, we consider the health benefits from reductions in emissions of local air pollutants along with the climate-related benefits from reduced CO_2.

The four main policy types differ in structure, in reliance on market forces, and in coverage across industries or fuels. They span many of the key dimensions that are the focus of climate-change policy discussions. We wish to reveal what is at stake in the choice among these options. The impacts depend on both the general policy approach and the specific design of a given approach. Indeed, a key theme that emerges from our analyses is that the particular design of any of the four policy types often is more important to overall cost and fairness than the choice among the four options. For example, under a carbon tax a key design feature is how the revenues from the tax are used. Depending on the method of revenue recycling, the costs will be higher or lower than the costs of a comparably scaled CES. In addition, the choice of recycling method for a carbon tax has a fundamental impact on how its impacts are distributed across firms, household income groups, and parts of the country. Design features are critically important to the cap-and-trade, gasoline tax, and CES policies as well.

This book offers answers to a range of questions relevant to the choice among the policy options:

- How much does it cost the overall economy to achieve given targets for reducing greenhouse gas (GHG) emissions? How do the costs compare with the climate-related benefits?
- What are the impacts on investment, household consumption, and GDP?
- What are the impacts on the profits of various industries in the U.S.?
- How are the costs distributed across household income groups and regions of the country?
- What does it cost to avoid disproportionate profit impacts on particular industries, or potential adverse economic impacts on low-income households?
- How do all these impacts change over time?
- How do the impacts depend on specific features of the policies, such as the stringency of the emissions reduction targets, the breadth of the policy (range of industries covered), and the way that policy-generated revenues are returned (or "recycled") to the private sector?

Although our analysis focuses on policies introduced in the U.S. and at the federal level, many of the lessons apply to other nations and to U.S. states looking to reduce CO_2 emissions cost-effectively and equitably.

THE MODEL

This book reports and interprets results from the Goulder-Hafstead Environment-Energy-Economy (E3) model, a multiperiod, economywide, general equilibrium model of the U.S. designed to answer these questions. The model is *multiperiod* in that it captures the path of the economy over time, enabling us to view the economic adjustments over many years into the future. It is *economywide* in that it embraces the entire U.S. economy. (It also considers foreign trade.) It is *general equilibrium* in nature in that it considers, in a consistent fashion, the interactions among various factor (labor and capital) markets and goods markets, and the interactions among key economic actors (the producing industries, the household sector, and the government). The model generates paths of equilibrium prices, outputs of goods and services, and incomes for the U.S. and the rest of the world under specified policies. It solves for all variables in each year, beginning with 2013.

While a number of other models possess the characteristics just described, the model used here combines two features that make it especially well suited to an analysis of U.S. climate-change policy options and distinguish it from other models.[9]

First, it considers important interactions between climate-change policies and the fiscal system. This is made possible by the model's combination of detail on the industries that supply or intensively use carbon-intensive fuels with detail on the U.S. tax system. A key theme of this book—supported by economic theory and reinforced by our numerical results—is that these interactions fundamentally affect the policy outcomes. Environmental policies interact with the tax system in three important ways. One is that the economic cost of every climate policy we consider depends intimately on the extent to which the economy already faces income, sales, or payroll taxes. A second is that the costs depend intimately on what is done with any policy-generated revenues. We will show that judicious revenue recycling substantially reduces policy costs and that the rankings of the four main policy types depend on the recycling method. A third form of interaction is that climate policies can affect incomes and thus the amount of revenue that existing taxes can generate. All three interactions play a major role in determining the policy outcomes described in later chapters.

A second important feature of the model is its attention to the dynamics of investment and disinvestment in physical capital, including the adjustment costs associated with the installation or removal of structures or equipment. This is especially important for ascertaining the impacts of policies on industry profits. Some models assume that physical capital is perfectly mobile across the economy, which means that capital can instantly flow out of one sector and into another in response to a change in economic conditions. This precludes an assessment of the differing profit impacts across industries, since it prevents consideration of stranded assets and instead implies that capital in all sectors instantly attains the same productivity following a policy intervention. Our model's recognition of the adjustment costs associated with the installation or removal of capital enables it to account for the windfall gains or losses in particular industries that occur under various climate policies, along with impacts on industry profits.

SOME KEY FINDINGS

The combination of attention to fiscal interactions and capital dynamics gives the model exceptional capabilities for revealing the differing impacts of important U.S. policy options. Key insights from our policy simulations include:

- If the environmental benefits are ignored, in most (but not all) cases the policies entail costs to the U.S. economy.
- However, at a scale large enough to achieve greenhouse gas emissions reductions similar in scale to what the U.S. had pledged under the 2015 Paris Agreement, the carbon tax, cap-and-trade, and CES policies produce climate-related benefits that exceed their costs by a significant amount. In contrast, reasonable increases in the federal gasoline tax do not generate emissions reductions of a magnitude comparable to those under the U.S. government's Paris pledge, and in some cases this policy's costs exceed its climate-related benefits.
- All policies generate co-benefits, mainly in the form of improved health from reduced air pollution. These typically exceed the climate benefits.
- For policies that raise revenues, recycling the revenues through cuts in the rates of preexisting taxes (such as payroll, individual income, and corporate income taxes) substantially lowers the policy costs, relative to the case when revenues are returned in a lump-sum fashion (as when revenues are returned through fixed rebates).
- In the absence of components offering targeted compensation, each of the climate policies produces very uneven profit impacts across

industries as well as adverse impacts on low-income household groups.

- However, at relatively little cost to the overall economy, both the adverse profit impacts on the most vulnerable industries and the negative welfare impacts for low-income household groups can be overcome.

- Emissions pricing through a broad-based carbon tax or cap-and-trade system is particularly attractive. When the revenues from a carbon tax or revenue-raising cap-and-trade system (one in which emissions allowances are auctioned) are recycled judiciously to the private sector, emissions reductions can be achieved at relatively low cost, and adverse distributional impacts across industries and households can be avoided.

ORGANIZATION OF THIS BOOK

The rest of this book is organized as follows. Chapter 2 completes part 1 by offering a conceptual background for understanding the economic forces behind the policy outcomes revealed later in this book. It indicates the ways in which the impacts of climate-change policies are shaped by interactions between these policies and the fiscal system. Part 2 includes chapter 3, which describes the structure of the numerical model, and chapter 4, which conveys the data and parameters used as inputs to the model. Part 3 presents and compares results from the various climate-change policies: a revenue-neutral carbon tax, a cap-and-trade system, a clean energy standard, and an increase in the federal gasoline tax. We explore both the economywide impacts (chapters 5 and 6) and the impacts across industries and households (chapter 7). Part 4 concludes, offering chapter 8, which summarizes our findings and indicates what our work suggests as promising directions for U.S. climate-change policy.

2

CLIMATE POLICY, FISCAL INTERACTIONS, AND ECONOMIC OUTCOMES

Proposed climate-change policies may differ in a number of ways, but they have at least one thing in common: they all interact with the fiscal system. And as we will show with the different climate-change policies considered in this book, the economic impacts of climate policies depend crucially on these interactions.

Perhaps the most obvious fiscal interaction relates to the use of policy-generated revenues. Several types of climate policy—including carbon taxes, increased gasoline taxes, and cap-and-trade systems in which the emissions allowances are auctioned—raise revenues. These revenues can be returned (or "recycled") to the private sector in various ways, including cuts in the rates of existing taxes such as individual income, corporate income, sales, and payroll taxes. Thus, a key potential interaction is the impact of climate policy on existing tax rates. As discussed later, recycling the revenues through cuts in tax rates can reduce policy costs. This cost-reducing impact is termed the *revenue-recycling effect*. The size of this effect depends on the particular tax that is lowered. An alternative to recycling through cuts in tax rates is to return the revenues in a lump-sum fashion, that is, through fixed rebates. In this case, there is no revenue-recycling effect.

A second interaction is the potential impact of climate policy on the revenue that existing taxes can collect, apart from any effect from changes to the rates of those taxes. By influencing incomes and spending, climate policies alter the base of other taxes and thereby affect the revenue that these other taxes yield. This is the *tax-base effect*. All the climate-change policies considered in this book have revenue implications through their impacts on the bases of the income tax and other existing taxes. Even the clean energy standard (CES), which is not a tax, has revenue impacts through its influence on incomes and the revenue collected by the income tax. For policies that aim to raise revenues—such as a carbon tax, a gasoline tax, or a cap-and-trade program involving auctioning of the emissions allowances—the total amount of revenue raised will depend on the tax-base effect. To the extent that climate policies come at a cost,[1] they reduce the bases of preexisting taxes and generate a gap between the policy's gross revenue and its net revenue—the revenue net of any policy-induced reductions in the revenues that other taxes collect. The larger the reduction in the tax base, the lower the net revenue generated, and thus the smaller the potential for reducing preexisting taxes and engaging the revenue-recycling effect.

A third interaction is between climate policies, the price level, and the real returns to factors of production (labor and capital). Although less obvious than the two previously described interactions, it is equally important and leads to what is termed a *tax-interaction effect*. By influencing the prices of goods and services relative to the prices of factors, climate policies affect the real returns to these factors. For example, the purchasing power of a person's wages falls when the prices of goods and services purchased by that person go up.[2] To the extent that they lower the real returns, climate policies function as implicit taxes on productive factors. This impact is termed a *tax*-interaction effect because its scale depends on the magnitudes of preexisting labor and capital taxes. In general, the tax-interaction effect works in the opposite direction from the revenue-recycling effect, tending to raise the costs of climate policies.

These fiscal interactions substantially affect the costs of climate policies, both in absolute terms and relative to one another. Choices about the design of a particular policy influence the magnitude of these interactions. Consider, for example, the case of cap and trade. Under this policy, a key design decision is whether to introduce the emissions allowances through free allocation or by way of an auction. In the case of free allocation, the policy generates no revenues, and there is no possibility of recycling through cuts in preexisting taxes. Hence there is no revenue-recycling effect. In the case of auctioning, the policy does generate revenues, and there is a revenue-recycling effect if the revenues are recycled through cuts in preexisting taxes. As we indicate later, this choice in policy design significantly affects the costs (and net benefits) of cap and trade and how its costs rank relative to those of other policies. Depending on the policy's design, a cap-and-trade program can be more or less cost-effective than the CES (where relative cost-effectiveness is the relative cost of achieving a given policy target, such as a given reduction in emissions or concentrations of CO_2).

Under all of the policies considered in this book, design choices determine the scope of fiscal interactions and thereby influence policy costs. Indeed, a key theme stemming from the explorations in this book is that design choices are at least as important as the choice among the various general policy instruments or types.

This chapter elucidates the nature of climate policies' fiscal interactions and reveals how these interactions shape the policy impacts. In doing so, it provides a conceptual basis for understanding the policy simulation results offered in later chapters. We offer a relatively nontechnical presentation here; a more technical and rigorous presentation can be found in appendix A.

ECONOMICS OF THE REVENUE-RECYCLING AND TAX-INTERACTION EFFECTS

To illustrate the importance of fiscal interactions, we start by focusing on the policy of a revenue-neutral carbon tax. We will indicate how the revenue-recycling and tax-interaction effects influence the overall costs of this policy (with the tax-base effect influencing the potential scope of the revenue-recycling effect). Later in the chapter we describe how these two effects apply to the other climate policies.

Consider first the basic economics textbook diagram offered in figure 2.1. This figure uses coal as an example of a commodity whose production generates a significant external cost. Hence the marginal social cost (MC_{SOC}) exceeds the marginal private cost (MC) by an amount reflecting the marginal external cost or damage (MED). Since our focus is climate-change policy, we will concentrate on the externality associated with the release of CO_2 from coal combustion and the associated impacts on climate and well-being. Of course, other externalities are associated

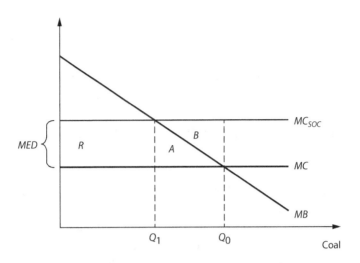

FIGURE 2.1

Simple Analysis of Benefits and Costs of an Environmental Tax

with coal combustion, including adverse health effects from emissions of local pollutants such as nitrogen oxide and sulfur dioxide.

In the absence of intervention, and assuming a competitive market for coal, the market equilibrium is at Q_0, where the marginal social cost from production exceeds the marginal benefit (MB) from the use of coal. The typical prescription from economists is to put a tax on coal that reflects the marginal external damage. We will refer to this as a carbon tax, implicitly assuming fixed proportions between the quantity of coal and the carbon content.[3] In a competitive market, this tax would cause private costs to rise and match the marginal social costs and would lead to the new equilibrium at Q_1, where the marginal social costs and marginal benefits are equal.

This policy yields an environmental benefit of $A + B$ (the avoided environmental damage). The diagram also indicates that this will cost society something. In particular, society will lose consumer surplus represented by $A + R$, as consumers (e.g., firms that generate electricity from coal) will have to pay more for coal. However, the tax revenue R confers benefits. If the revenue is recycled back to the economy (rebated to the private sector), the private sector gets back the value R.

Alternatively, it could be used to finance new government spending. If the benefit-cost ratio for the financed project is about 1, this spending has a value of about R. Thus, the textbook diagram suggests that, ignoring environmental benefits, the overall cost to the economy is A. This policy offers net social benefits of B, since the environmental benefit $A + B$ exceeds the cost A.

The analysis presented thus far ignores the fiscal interactions described earlier. In the early 1990s, a number of economists began to consider these interactions. Many of their analyses suggested that the simple textbook framework understated the social benefits from the tax policy. They emphasized that when tax revenue R is returned to the private sector, it can have a social value greater than R. They pointed out that, rather than provide for lump-sum payments, the carbon tax revenue could be used to finance reductions in the marginal rates of existing distortionary taxes such as income or sales taxes. Since the distortionary cost (deadweight loss or excess burden) of these taxes is a positive function of their tax rates, lowering these tax rates could reduce the distortionary cost of these taxes. The revenue-recycling effect referred to earlier is the efficiency improvement from cutting these tax rates.[4] The simple textbook analysis ignores this positive efficiency effect.

The net revenue from the carbon tax is the gross revenue adjusted for the tax base effect, that is, for any change in the revenue earned from existing taxes prior to any change in their tax rates. Throughout this book we will refer to a *revenue-neutral* policy as one in which all of the net revenue is recycled (returned) to the private sector. To the extent that an adverse tax-base effect reduces net revenue, it will limit the extent of recycling for a revenue-neutral policy and thereby reduce the scale of the revenue-recycling effect. Although the policies explored in later chapters do produce an adverse tax-base effect, which reduces the magnitude of the revenue-recycling effect, the revenue-recycling effect remains large enough to have a profound impact on policy costs.[5]

Attention to the (beneficial) revenue-recycling effect caused some to suggest that this environmental policy might be a zero- or negative-cost option, even before accounting for the environmental benefits. That is, the efficiency improvement from revenue recycling might fully offset the cost A that applies before this efficiency component is considered. If so, the tax policy would confer a *double dividend*—an environmental benefit *and* higher nonenvironmental well-being by lowering the cost to the private sector of the tax system.[6] Environmental improvement could be a free lunch. For advocates of environmental taxes, the possibility of a double dividend—a zero-cost option—was quite appealing.

But subsequently some other economists began to cast doubt on these relatively optimistic conclusions. They drew attention to another effect that works in the opposite direction, magnifying some of the costs of the environmental tax. Bovenberg and de Mooji (1994), for example, argued that "environmental taxes typically exacerbate, rather than alleviate, preexisting tax distortions—even if revenues are employed to cut preexisting distortionary taxes." These authors, as well as Parry (1995) and Goulder (1995), emphasized the tax-interaction effect introduced earlier, which reflects the fact that environmental taxes, like all taxes on goods and services, are implicit taxes on factors of production.[7]

The tax-interaction effect arises to the extent that preexisting taxes (such as income, payroll, or sales taxes) have produced distortions in factor markets. Figure 2.2 illustrates the issues, focusing on the labor market.

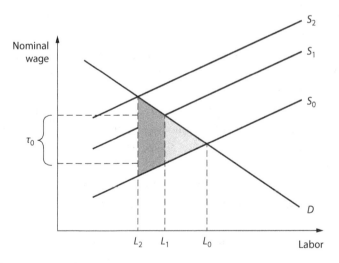

FIGURE 2.2

The Tax-Interaction Effect

Existing taxes impose a wedge between the marginal value of labor supplied and the private marginal cost of labor supply. In the figure, τ_0 represents the tax wedge implied by these existing taxes. With this tax wedge, the market equilibrium labor supply is at L_1, below the most efficient level L_0. The excess burden associated with these existing taxes is given by the light shaded triangle in the diagram.

Now suppose that, in this situation, a carbon (coal) tax is introduced. This implicit tax on labor raises the labor supply curve further, to S_2. This leads to a further reduction in labor supply; the new equilibrium labor supply is L_2. Associated with this further reduction is an additional excess burden or efficiency loss, represented by the dark shaded trapezoidal area in the diagram. This additional efficiency loss—stemming from the carbon tax's functioning as an implicit tax on labor (or, more generally, on factors of production)—is the tax-interaction effect. Note that the higher the preexisting taxes on factors (or, in the diagram, on labor), the more pronounced is the tax-interaction effect.

While demonstrated through a focus on a carbon tax, the tax-interaction, tax-base, and revenue-recycling effects apply to the other climate policies considered in this text as well. In fact, any policy that raises prices of goods and services will produce a tax-interaction effect, any policy that affects incomes has a tax-base effect, and any policy whose revenues are returned to the private sector through cuts in prior taxes exerts a revenue-recycling effect.

WHAT'S THE OVERALL IMPACT AFTER ACCOUNTING FOR THESE INTERACTIONS?

An Initial Look

This discussion has identified three main determinants of the costs of an environmental tax: the primary cost (area A in figure 2.1), the tax-interaction effect, and the revenue-recycling effect (whose magnitude is influenced by the tax-base effect). Whether the overall cost exceeds the primary cost depends on the relative magnitudes of the tax-interaction effect (which adds to cost) and the revenue-recycling effect (which reduces cost).

The literature on environmental taxation over the past two decades indicates that the tax-interaction effect often dominates the revenue-recycling effect. The intuitive explanation for this focuses on the differences in the "distortions" created by the environmental tax and the "distortions" removed through the revenue-recycling effect. (We put "distortions" in quotes to acknowledge that this change may be beneficial in terms of efficiency once environmental impacts are taken into account.) The environmental tax distorts not only the factor market (by lowering the returns to factors) but also the patterns of consumption by altering the relative prices of goods consumed. The recycling of revenues through cuts in income or sales tax rates can offset the former distortion, but the change in relative prices of goods consumed remains. Although this change in relative prices is attractive on environmental grounds—indeed, it supplies the key motivation for an environmental tax!—it is irrelevant to the assessment of the second (nonenvironmental) dividend. Thus, the revenue-recycling effect is not able to offset the tax-interaction effect.

To view this in greater detail, consider the following simple economic model. Assume a one-period closed economy model with one factor of production—labor—and a preexisting tax on that factor—a labor income tax. Consider first the impact of introducing in this setting at the margin a new tax, a uniform tax on consumer goods. Let the revenue from this new tax be recycled in the form of equal-revenue cuts in the preexisting labor tax. In this case, the tax-interaction and revenue-recycling effects cancel. The reason is that in this simple model a consumption tax and a labor tax are equivalent. The incremental consumption tax works to lower the real wage by raising P, the price index for goods and services. At the same time, revenue recycling lowers the explicit tax on labor, which works to raise the real wage by increasing the after-tax nominal wage W. The two effects cancel: the real wage is unaffected. Hence this revenue-neutral tax swap doesn't change labor supply or involve any cost to the economy. Indeed, it has no economic or environmental impact. It doesn't cost the economy anything, but it doesn't yield any environmental improvement either.

Now consider an alternative revenue-neutral tax policy, one that is more applicable to environmental protection. In the same setting as before, consider the impact of introducing an environmentally motivated tax on a particular consumer good (e.g., gasoline), rather than the broad-based commodity or income tax. As before, assume that revenues are recycled through cuts in the existing labor tax. As in the previous case, the policy is revenue-neutral, but in this case the new tax is on a specific good.

In this case the tax-interaction and revenue-recycling effects do not cancel. As before, the new tax raises P while revenue recycling raises the after-tax nominal wage W. But here a new "distortion" is produced. (Again we put "distortion" in quotes because it's not a distortion if one accounts for the environmental benefits—but right now we're considering whether there's a cost in the other, nonenvironmental, dimension.) There is an extra cost in this case because the gasoline tax "distorts" the price of gasoline relative to other commodities. If we ignore the environment, this hurts efficiency. The revenue-recycling effect does not entirely offset the tax-interaction effect. Putting aside the environmental benefits, the revenue-neutral tax swap makes the tax system more costly because it introduces a cross-commodity distortion.[8]

These ideas are embodied in figure 2.3, which displays connections between the primary cost, the tax-interaction effect, and the revenue-recycling effect. Appendix A offers a more rigorous mathematical

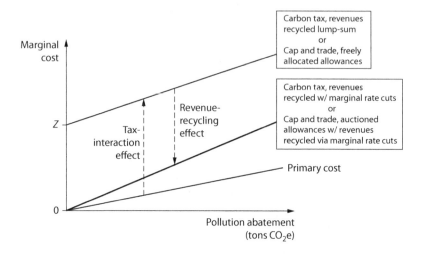

FIGURE 2.3

Marginal Costs of Pollution Abatement with General Equilibrium Effects presentation.[9] The horizontal axis is the extent of CO_2 abatement. Think of abatement as a function of the carbon tax rate. As you move from left to right in the figure, the increased abatement reflects an increase in the price of emissions.

Primary cost (area A in figure 2.1) increases with abatement; this assumes that firms' marginal abatement costs are positive.[10] The tax-interaction effect also increases with abatement. The line representing this effect has a positive intercept: given positive preexisting factor taxes, the first unit of abatement produces a strictly positive tax-interaction effect. That is, even a one-unit decrease in emissions will reduce real factor incomes, increasing the cost of emissions reductions.[11] When revenues from the emissions pricing policy are recycled through cuts in the rates of preexisting taxes, the revenue-recycling effect kicks in. The darker black line represents the overall cost at different levels of abatement: primary cost plus the tax-interaction effect minus the revenue-recycling effect. As drawn, the overall cost line lies above primary cost, in keeping with the idea that the revenue-recycling effect does not completely offset that tax-interaction effect.

Exploiting the revenue-recycling effect not only is important for cost-effectiveness but also can determine whether emissions pricing achieves an increase in efficiency, that is, whether it yields (monetized) environmental benefits in excess of the costs. In figure 2.3, the point marked by Z is the intercept of the line for the tax-interaction effect. The height of this intercept represents the magnitude of the tax-interaction effect for an initial amount of CO_2 abatement.[12] This number is important because it implies a critical value for environmental benefits. If the environmental benefit per ton at the margin from reducing emissions is less than Z, then any amount of reduction in emissions from the carbon tax will involve costs greater than the benefits if the revenue-recycling effect is not exploited (assuming marginal environmental benefits are constant or increasing). This effect is not engaged when recycling is done in lump-sum fashion, such as when revenues are returned through rebate checks to households.

To get the double dividend, the revenue-recycling effect would need to outweigh both the tax-interaction effect and the primary cost. In that case, the dark line in figure 2.3 would be below the horizontal axis. Our original analysis casts doubt on this result.

A Second Look: A Double Dividend After All?

For some advocates of revenue-neutral carbon taxes or other environmental tax policies, the absence of a double dividend would not be a welcome result. It would be much easier to promote such policies if their gross costs—the costs before accounting for environmental benefits—were zero or negative.

However, certain additional factors could cause the revenue-recycling effect to outweigh the combination of the tax-interaction effect and the primary cost, which would allow the double dividend to occur after all. Table 2.1 lists some circumstances conducive to the double dividend. In the first four circumstances listed, the potentially low cost of the revenue-neutral environmental tax stems from two conditions: (1) the preexisting tax system was inefficient along some *nonenvironmental* dimension, and (2) the revenue-neutral environmental tax policy served to reduce this nonenvironmental inefficiency. The last two circumstances can yield low costs because of a direct connection between an environmental improvement and utility from consumption or the productivity of factor inputs.

The first circumstance pertains to the relative taxation of capital and labor. In some economies there is relatively inefficient relative taxation of these factors. Studies of the U.S. tax system provide evidence that the marginal excess of burden of taxes on capital is significantly higher than that of labor taxes.[13] In such a setting, if the government introduces

TABLE 2.1 **Circumstances Conducive to the Double Dividend**

Nonenvironmental inefficiencies in the tax system:
- Inefficient relative taxation of capital and labor
- Inefficiently light taxation of resource rents
- An informal labor market and associated inefficiently low taxation of informal labor income
- Income-related distorting subsidies to consumption goods and services

Other:
- Complementarities between environmental quality and marketed goods or services
- Positive relationship between environmental quality and labor productivity

an environmental tax—a carbon tax or gasoline tax, for example—and devotes the revenue to cutting capital taxes, then the discrepancy is reduced. If the associated nonenvironmental improvement in the efficiency of the tax system is large enough, the revenue-recycling impact of reducing capital taxes could outweigh both the primary costs and the tax-interaction costs, in which case the double dividend would be realized.[14]

The second circumstance listed refers to the relative taxation of resource rents and other factors. In analyzing the prospects for a double dividend, Bento and Jacobsen (2007) start with the observation that a significant share of the burden of a carbon tax falls on resource rents.[15] When a carbon tax is introduced and its revenues are used to finance cuts in labor taxes, the tax burden is shifted away from labor toward resource rents. Since the resources yielding these rents are inelastically supplied (that is why they generate rents), this shift works toward an improvement in the efficiency of the tax system. If the associated efficiency gain is large enough, the costs of the tax system can be reduced: the double dividend is obtained.

A third situation relates to informal labor markets. Because the informal sector faces no labor tax, there is a preexisting distortion in the economy: labor in the informal sector is undertaxed relative to labor in the formal sector. In contrast with the labor tax, a carbon tax is a tax on both forms of labor because it raises the prices of goods purchased by both forms of labor. It thereby lowers the real wage to both forms of labor. When the revenues from the carbon tax are recycled through cuts in the tax on formal labor, the net effect is to reduce the discrepancy in the taxation of formal and informal labor, which improves efficiency. If the associated efficiency gain is large enough, the costs of the tax system can be reduced. Moreover, overall employment can increase.[16] In the U.S., the informal labor market is small relative to the formal market. However, in many developing countries the informal market has greater importance and could offer significant opportunities for the double dividend.[17]

The fourth situation refers to preexisting subsidies that can distort the allocation of consumption expenditure across various goods and services. In the U.S. tax system, taxpayers enjoy a range of tax deductions for particular expenditures, including home purchases and medical insurance.

The higher the taxpayer's income tax rate at the margin, the greater the value of the deduction to the taxpayer. Thus, when the revenues from an environmental tax are recycled in the form of cuts in marginal income tax rates, the value of these deductions is reduced. To the extent that the deductions are distortionary (i.e., efficiency-reducing), the recycling via cuts in income taxes works to improve efficiency. Parry and Bento (2000) suggest that revenue-neutral environmental taxes are often strong enough to yield a double dividend as a result of the beneficial efficiency impacts associated with such revenue recycling. It is worth noting, however, that this argument applies only to the extent that the tax deductions are distortionary. Some would argue that for some types of consumer expenditure (e.g., home ownership) the favorable tax treatment helps internalize external benefits.[18]

The final two circumstances listed in the table differ from the first four in that they do not reflect a preexisting inefficiency in the tax system on nonenvironmental grounds, but rather offer a channel for additional benefits from a revenue-neutral environmental tax beyond the environmental (e.g., climate) improvement itself.[19] If we view these additional benefits as negative costs, the overall costs of the revenue-neutral environmental tax policy are reduced through these channels and can be negative.

The fifth situation is one in which environmental quality and some marketed goods are complements in utility, meaning that an improvement in environmental quality raises the marginal utility from consumption of a given quantity of a marketed good. This increase in the marginal utility from consumption (caused by preventing decreases in environmental quality) can be interpreted as a reduction in the cost of the revenue-neutral policy.[20]

The sixth situation is one in which there is a direct and positive connection between environmental quality and productivity. For example, with climate change, workers may be less productive when it is hotter, agricultural yields may fall due to increased weather variability, and fishermen may find it harder to produce target catch sizes owing to adverse impacts of warmer oceans.[21] Preventing these decreases in productivity by reducing emissions that lead to climate change yields an additional benefit of revenue-neutral environmental tax policy which, when viewed as a negative cost, can cause overall costs to become negative. In later chapters, we will explore the potential for a double dividend numerically, using our general equilibrium model.

The possibility of zero or negative costs has gained considerable attention. It is important, however, not to lose sight of another general notion that is equally or more significant: that exploiting the revenue-recycling effect can substantially lower a policy's costs, regardless of whether costs are brought below zero. We address this issue later in this chapter.

FISCAL INTERACTIONS FROM THE OTHER CLIMATE POLICIES

Fiscal interactions—notably, the tax-interaction and revenue-recycling effects—also arise under the other climate policies considered in this text. And, as under the carbon tax, they significantly influence the economic outcomes of these other policies.

Cap and Trade

Under cap and trade, the revenue-recycling effect depends on whether the policy generates revenue and on how such revenue is recycled to the private sector. Figure 2.3, which we examined in connection with the carbon tax, also reveals key issues under cap and trade. Consider first a cap-and-trade system in which all the allowances are auctioned and the revenues from the auction are used to finance cuts in the marginal rates of preexisting taxes. As with a carbon tax, cap and trade introduces a tax-interaction effect, since it raises the prices of goods and services in general and thereby serves to lower real factor returns, abstracting from revenue recycling. If the auction revenues are used to finance cuts in prior taxes, the policy also yields a revenue-recycling effect, which helps contain the policy costs. The dark black line represents the overall cost of the policy.

A cap-and-trade system with a different design would involve very different policy costs. If all the allowances are auctioned but the net revenues are recycled in lump-sum fashion (e.g., through fixed-value rebates to households), then there is no revenue-recycling effect and the top line represents the overall costs. If all the allowances are given out free rather than auctioned, the net revenue is negative if there is an adverse impact on the tax base, since there is no gross revenue to off-set the adverse tax-base effect. A revenue-neutral cap-and-trade system with free allocation of allowances therefore tends to require *increases* in preexisting tax rates. In this case, the overall costs of the policy are greater than the top line in figure 2.3.

As in the case of the carbon tax, the presence or absence of the revenue-recycling effect can determine whether the policy achieves an increase in efficiency. The intercept Z in figure 2.3 again applies. If the revenue-recycling effect is absent (or negative) and the monetized environmental benefits per ton reduced are less than the height of this intercept, then emissions pricing will not yield net benefits. However, if the cap-and-trade system involves recycling of the auction revenues through cuts in preexisting taxes, then the revenue-recycling effect applies. For any given level of environmental benefits, the range of abatement levels over which net benefits are positive is wider when the revenue-recycling effect is pres-ent than when it is absent. Given the substantial uncertainty in the value of the environmental benefits from preventing climate change, cap-and-trade programs that utilize the revenue-recycling effect are more likely to deliver net benefits than policies that do not. We explore these issues numerically in chapter 5.

The Clean Energy Standard

A CES yields a tax-interaction effect through its impact on the prices of goods and services. But because it yields no gross revenue, it cannot produce an offsetting revenue-recycling effect. This might suggest that the overall impact of fiscal interactions is more negative under the CES than under the other climate policies considered. However, as explored in chapter 6, the CES generally produces a smaller tax-interaction effect than an equally stringent cap-and-trade or carbon tax policy would. This reflects the fact that the CES tends to raise electricity prices (and hence the prices of goods and services overall) by less than the other policies do. The smaller tax-interaction effect can compensate for other disad-vantages of the CES in terms of cost-effectiveness. In fact, as chapter 6 indicates, when the target for emissions reductions is modest, the target can sometimes be reached at lower cost under a CES than under cap and trade or a carbon tax.

An Increase in the Federal Gasoline Tax

As with the other climate policies, an increase in the federal gasoline tax generally produces tax-interaction and revenue-recycling effects. For the gasoline tax increase, the tax-interaction effect can be significant. Empirical studies find that the price elasticity of demand for gasoline is fairly low—well below 1 in the short and medium run.[22] A low elasticity of demand implies that a gasoline tax could occasion significant increases in gas prices per unit of tax, implying a large tax-interaction effect. As with the other policies, the costs will depend crucially on how the policy-generated revenues are recycled to the private sector.

The Substantial Impact of Marginal Rate Cuts on Policy Costs

The cost difference between exploiting and forgoing the revenue-recycling effect can be very large. We explore this issue in later chapters, but it seems worthwhile to acknowledge here some prior assessments that relate to it. Parry and Williams (2010) evaluate the impacts of a cap-and-trade system to reduce CO_2 emission by about 9 percent by 2020. They estimate the GDP cost to be about $53 billion per year when the allowances are given out free (and there is no revenue-recycling effect); the GDP cost vanishes when the allowances are auctioned and the revenues are devoted to cuts in income tax rates. Similarly, Goulder and Hafstead (2013) find that, for a carbon tax rising from $10 to $37 per ton from 2013 to 2040, the GDP costs over that interval are on average 41 percent lower when the revenue is used to reduce personal income tax cuts and 56 percent lower when corporate income tax rates are reduced.

Similar cost differences apply under other climate policies. Under cap and trade, the revenue-recycling effect is engaged when emissions allowances are introduced through an auction and the revenues obtained are recycled through cuts in marginal tax rates. In contrast, the revenue-recycling effect is absent when auction revenues are returned in lump-sum fashion. And as discussed earlier, this effect is negative when allowances are introduced through free allocation. Goulder, Hafstead, and Dworsky (2010) indicate that the GDP costs of reducing U.S. CO_2 emissions under cap and trade are approximately one-third lower with auctioning and recycling via personal income tax cuts than under free allowance allocation.

Trade-offs between Cost-Effectiveness and Distributional Equity

While exploiting the revenue-recycling effect has advantages in terms of cost-effectiveness, taking full advantage of this effect might not address concerns about distributional equity. Compared with more affluent households, low-income households tend to spend a larger share of their incomes on carbon-intensive goods and services such as home heating and transportation. Since climate policies tend to raise the relative prices of carbon-intensive goods and services, they can have a regressive impact. And revenue recycling through general tax cuts may reinforce the regressive impacts, especially if carbon revenues are used to reduce capital taxes, since a majority of capital is owned by the upper end of the income distribution.

Revenue recycling need not have a regressive impact, however. Regressive outcomes can be reduced or avoided by returning revenues through fixed payments (rebates) of equal amounts to households, rather than through cuts in marginal tax rates. Equal rebates to households at all levels of income can reduce the regressive impact since, relative to income, the rebate will be larger for the lower-income households.

Many policy analysts and decision makers, from both the left and the right side of the political spectrum, have favored this approach.[23] The progressive Citizens' Climate Lobby has long promoted a "carbon fee and dividend" (carbon tax with lump-sum rebates). The Climate Leadership Council, led by eminent conservative policy makers such as James A. Baker, III, George P. Shultz, Martin Feldstein, and N. Gregory Mankiw, also advocates for lump-sum dividends.[24] In addition, this approach has been adopted as part of some carbon tax policies currently in place.[25]

However, there is a trade-off here. As indicated by this analysis, this method of recycling comes at a cost in terms of efficiency because in this case the revenue-recycling effect is smaller than that when all the revenues are recycled through tax cuts. In Chapter 7 we apply our E3 model to assess this trade-off numerically.

Concerns about distributional equity also arise in connection with uneven impacts of climate policies across industries or geographic regions. Many politicians have expressed particular concern about the potential impact of climate policies on the coal industry, which could suffer a disproportionate adverse impact. States with considerable coal mining, such as West Virginia, Kentucky, and Wyoming, could potentially suffer large losses of profit and employment relative to other states. Again, revenue recycling has the potential to offset or even eliminate the adverse impacts. As explored quantitatively in chapter 7, profit losses can be avoided by using some of the policy-generated revenue to compensate especially vulnerable industries. However, doing so would leave less revenue for recycling in the form of general tax cuts. Once again, a trade-off arises between cost-effectiveness and distributional equity (and political feasibility).

How to strike a balance between these competing goals is a key challenge to policy makers. Our analysis cannot determine where that balance is best struck: the difficult ethical questions raised here are beyond the scope of this book. But we can contribute to policy discussions by revealing the nature and scope of the trade-offs. That is, we can indicate the sacrifices of cost-effectiveness associated with a range of policies that address distributional concerns in various ways and to differing extents. We delineate these equity-efficiency trade-offs in later chapters.

SUMMARY

Climate-change policies interact with the fiscal system through tax-interaction and revenue-recycling effects as well as their impacts on tax bases. This chapter conveys the economics behind these effects and reveals how they influence the economywide costs of climate policies and the distribution of their impacts.

A key choice by policy makers is how much to exploit the revenue-recycling effect, that is, how much to recycle the revenues through cuts in the marginal rates of preexisting taxes. This choice can determine whether the net benefits of a given policy are positive or negative. It can also influence the relative costs of different policies. Choosing which taxes to cut and by how much is as important as the choice of the policy instrument. For example, whether a carbon tax is more or less cost-effective than a clean energy standard depends on how the carbon tax revenues are recycled.

While exploiting the revenue-recycling effect has advantages in terms of cost-effectiveness, taking full advantage of this effect typically ignores concerns about distributional equity. Achieving a more even or equitable distribution of impacts across household income groups, industries, or regions often requires returning revenues in a form other than cuts in marginal rates. This implies a trade-off between cost-effectiveness and distributional equity. Combining a carbon tax with cuts in capital tax rates, for example, may be desirable from a cost-effectiveness perspective, but may yield a highly regressive outcome.

In later chapters we consider a range of policy experiments in which we examine quantitatively both the economywide and the distributional impacts of climate policies. We also evaluate trade-offs between cost-effectiveness and distributional equity. These examinations provide information that can help guide decision makers in their choices among different policy instruments and—equally important—their decisions about the design of given instruments.

II

THE MODEL'S STRUCTURE, INPUTS, AND BASELINE OUTPUT

3

STRUCTURE OF THE E3 MODEL

The E3 (energy-environment-economy) model is designed to convey the impacts of U.S. environmental and energy policies. As a general equilibrium model, it considers interactions among industries and markets. This contrasts with partial equilibrium models, which focus on a given industry or market in the economy. A key finding from the theoretical framework in chapter 2 and the numerical findings in later chapters is that these interactions fundamentally affect the outcomes of climate policies, in some cases giving results that differ dramatically from what a partial equilibrium analysis would suggest.

Computable general equilibrium (CGE) models combine economic data with micro-founded formulas for economic behavior to predict how included economic variables change over time.[1] The solution to these models is a set of prices that yields a *Walrasian general equilibrium*—a matching of the supplies and demands of all goods and factors in all periods.

Figure 3.1 conveys the interactions of various markets in a CGE framework. It gives an idea of the types of interactions in the E3 model by displaying the circular flow of goods, services, expenditures, and receipts in a simplified and closed economy. The key agents in both the figure and the E3 model are producers, households, and the government. Households provide labor and capital to the producers, who then provide goods and services to the households, other producers, and the government. Households pay for those goods and services with labor and capital income, where capital income includes the returns from ownership of firms as well as the returns on owned public and private bonds. The government

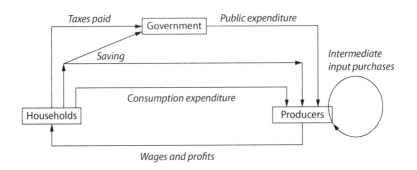

FIGURE 3.1

Flows of Goods and Payments

levies taxes on households and firms, issues bonds, transfers money to households, and purchases good and services.

As the figure indicates, the expenditures and receipts of all agents are linked: each expenditure by a given party corresponds to a receipt by another party. In addition, each agent's expenditures must be consistent with its receipts. Household expenditures on goods and services, plus household savings, cannot exceed after-tax income (household budget balance). The government's expenditures must not exceed its tax revenues plus new borrowing (the government budget constraint). For each firm, the payments to labor, materials, and other inputs, plus taxes paid and the funds distributed as profit or kept as retained earnings, must be equal to revenues from sales of the goods and services (firms' budget balance). In this system, the Walrasian general equilibrium occurs when the demands for goods and services and factors of production match the supplies of those goods and services and factors of production. This match is *market clearing*. Numerically, the E3 model finds the prices that bring about market clearing in each period of time.

Although CGE models have the basic framework in common, they can differ greatly in the specifics. The rest of this chapter describes the E3 model, giving particular attention to its distinguishing features. Here we offer an overview of the behavior of the model's key agents and indicate how market-clearing prices reconcile supplies and demands. Detailed descriptions of the model's structure and solution algorithm are offered in appendix B.

E3 MODEL KEY FEATURES

E3 is a multisector and multiperiod dynamic model of the U.S. economy with international trade. Agents are assumed to be forward-looking (that is, they take future prices into consideration in making current decisions) and to have perfect foresight. A perfect-foresight intertemporal equilibrium is achieved only when markets clear in every period *and* agents' expectations of future prices match the equilibrium prices of future periods. Each period of the model corresponds to one year.

As indicated in this text's introduction, two distinguishing features of the model are (1) its attention to interactions between climate policy and the fiscal system and (2) its consideration of the dynamics of investment and disinvestment in physical capital. To capture important impacts of climate policy, the model gives particular attention to energy industries and the relationships among them. The model's treatment of energy recognizes the nonrenewable aspect of oil supply, differences in the carbon intensity of various forms of electricity generation, and the gradual phase-in of "backstop" supplies of energy. On the tax side, the model distinguishes various forms of capital taxation (separately treating taxes on corporate profits, dividends, interest income, and capital gains), and distinguishes payroll taxes from other taxes on labor income. It also recognizes sales taxes and various tax preferences such as deductions for certain forms of consumer spending. The detailed treatment of both the energy system and the tax system allows the model to consider critical interactions between climate policy and the fiscal system. As emphasized in chapter 2, these interactions fundamentally shape the impacts of climate policies.

In considering investment dynamics, the model recognizes the adjustment costs associated with the installation or removal of physical capital. These costs affect the pace of the economy's transition to new production methods and consumption patterns in response to new environmental policies. They also are critical to revealing how a given climate policy can have very different effects on the profits and value of assets across industries. Indeed, models without adjustment costs (or, more generally, models with perfectly mobile physical capital) cannot account for differences across industries in the impacts of policies on profits.

PRODUCER BEHAVIOR

The model incorporates key channels through which producers respond to changing economic conditions. Producers alter the mix of variable inputs to minimize costs and make investment decisions aimed at maximizing the value of the firm.

Production in the U.S. is divided into thirty-five industry categories. This choice of categories gives focus to energy supply and demand. The crude oil extraction, natural gas extraction, and coal mining industries supply the fuels demanded by the electricity sector industries, natural gas distribution utilities, petroleum refiners, and other industries. Table 3.1 lists the industries, the value of output of each industry, and the energy and labor inputs for each industry.[2] The table also reports the relative size of each industry and the energy and labor intensity of each industry.

The model recognizes the nonrenewable nature of oil resources and the associated fact that extraction costs rise as reserves decline. Oil producers account for the impact of current production on future costs when making profit-maximizing extraction decisions.

The electric power sector is a major source of carbon emissions. Reductions in these emissions can be achieved in various ways, including shifting generation away from coal-fired plants toward natural gas-fired plants or shifting toward nonfossil electricity sources such as solar, wind, hydro, and nuclear power. We model electricity generation through three types of suppliers: coal-fired electricity generators, other fossil electricity generators, and nonfossil electricity generators.[3] This specification allows for changes in the generation mix in response to federal environmental policy. The electric utility industry in the model purchases electricity from the three different generators on the wholesale market and is responsible for transmitting and distributing retail electricity to households and industries. Retail electricity prices reflect both wholesale electricity prices and the costs of delivering electricity to customers.

Production

A representative firm in each industry produces distinct output X, using two categories of capital (K^s and K^e), labor L, energy intermediate inputs ($E_1, ..., E_{N_e}$), and nonenergy (or materials) intermediate inputs ($M_1, ..., M_{N_m}$).[4]

TABLE 3.1 Output and Energy Input by E3 Industry, 2013

INDUSTRY	OUTPUT[a]	PCT OF TOTAL OUTPUT	ENERGY INPUT[b]	PCT OF OUTPUT	LABOR INPUT[a]	PCT OF OUTPUT
Oil extraction	253.8	1.0	6.9	2.7	7.2	2.8
Natural gas extraction	110.3	0.4	2.9	2.7	3.2	2.9
Coal mining	40.7	0.2	2.5	6.1	4.1	10.0
Electric transmission and distribution	359.4	1.4	200.5	55.8	34.2	9.5
Coal-fired electricity generation	69.6	0.3	22.2	31.8	9.0	12.9
Other-fossil electricity generation	64.6	0.3	36.1	55.9	6.0	9.3
Nonfossil electricity generation	53.3	0.2	0.1	0.1	9.7	18.1
Natural gas distribution	126.1	0.5	47.4	37.6	15.5	12.3
Petroleum refining	694.5	2.7	552.4	79.6	11.8	1.7
Pipeline transportation	38.8	0.2	3.1	8.0	5.7	14.7
Mining support activities	196.3	0.8	5.8	2.9	56.2	28.6
Other mining	46.7	0.2	5.8	12.3	6.3	13.5
Farms, forestry, fishing	424.9	1.7	25.4	6.0	48.2	11.4
Water utilities	78.7	0.3	1.8	2.3	3.0	3.8
Construction	1,371.9	5.4	52.7	3.8	409.0	29.8
Wood products	92.1	0.4	2.8	3.0	16.4	17.8
Nonmetallic mineral products	103.8	0.4	6.0	5.8	21.5	20.7
Primary metals	286.4	1.1	18.1	6.3	28.1	9.8
Fabricated metal products	334.9	1.3	6.9	2.0	81.8	24.4
Machinery and miscellaneous manufacturing	1,358.0	5.4	12.5	0.9	309.1	22.8
Motor vehicles	581.5	2.3	4.3	0.7	50.4	8.7
Food and beverage	784.0	3.1	12.9	1.6	89.3	11.4
Textile, apparel, and leather	84.3	0.3	1.5	1.7	19.8	23.5
Paper and printing	227.4	0.9	11.9	5.2	52.1	22.9
Chemicals, plastics, and rubber	990.4	3.9	62.7	6.3	101.4	10.2
Trade	2,374.1	9.4	34.0	1.4	830.8	35.0
Air transportation	158.0	0.6	36.1	22.8	40.5	25.7
Railroad transportation	99.4	0.4	6.5	6.6	21.2	21.3
Water transportation	50.5	0.2	9.4	18.7	5.7	11.2
Truck transportation	282.7	1.1	49.7	17.6	75.3	26.6
Transit and ground passenger transportation	56.5	0.2	5.5	9.7	16.3	28.9
Other transportation and warehousing	286.8	1.1	16.2	5.6	102.1	35.6
Communication and information	1,145.7	4.5	4.9	0.4	241.3	21.1
Services	9,659.4	38.2	113.6	1.2	4,082.0	42.3
Real estate and owner-occupied housing	2,371.4	9.4	92.9	3.9	76.6	3.2
Total	**25,256.9**	**100.0**	**1473.6**	**5.8**	**6890.8**	**27.3**

[a] In billions of 2013 dollars.
[b] In billions of 2013 dollars. Energy inputs include fossil fuels, electric power sectors, natural gas distribution, and petroleum refining.

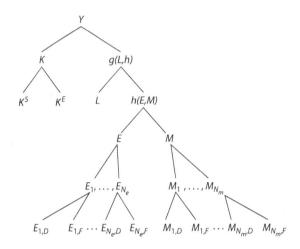

FIGURE 3.2

E3 Production Structure

Producers choose variable inputs to minimize costs and make investment decisions to maximize the value of the firm—the present value of after-tax dividends minus new share issues.

Output from each industry stems from a nested structure of constant-elasticity-of-substitution production functions. Figure 3.2 displays this structure. At each node, an aggregation function creates a composite of the inputs immediately below that node.[5] A total of forty-one functions are used for each industry.

At the lowest level of the nested structure, each input $E_1, ..., E_{N_e}$ and $M_1, ..., M_{N_m}$ is a composite of the good produced by the domestic industry and the good produced by its foreign counterpart (for example, $M_1 = d_{m1}(M_{1,d}, M_{1,f})$). One level up, intermediate inputs of energy and of materials are aggregated into an energy composite E and a materials composite M: $E - e(E_1, ..., E_{N_e})$ and $M = m(M_1, ..., M_{N_m})$.

The nested production structure then uses E, M, L, and K to produce the output X. This structure is represented by

$$X = f(K, g(L, h(E, M)))$$ (3.1)

As equation (3.1) indicates, the energy and materials composites E and M are aggregated via the function h into an intermediate input composite.

The function g then creates a composite from L and the composite h. The top-level function f next combines g and the capital composite K to yield output X. The capital composite K consists of structures K^s and equipment/intellectual property products K^e, $K = K(K^s, K^e)$. For brevity we will often refer to K^e simply as equipment, though in fact it captures intellectual property products as well.

The model includes an exogenously specified rate of technological change. We assume that such change is Harrod-neutral (labor-embodied). For any given mix of productive inputs (labor, capital, energy, and materials), the productivity of the labor input increases as a function of time. Thus, effective hours worked are actual hours adjusted for the rate of technological change. This rate of change will also dictate the growth of the quantities of all producer inputs and outputs on the steady-state balanced growth path.[6]

Investment

In each period, managers invest in structures and equipment, taking into account the adjustment costs associated with the installation or removal of physical capital. Capital adjustment costs are modeled as lost output associated with the process of investing in structures and equipment: net output is equal to gross output minus the adjustment costs:

$$Y = X - \phi(I^s/K^s) \cdot I^s - \phi(I^e, K^e) \cdot I^e \qquad (3.2)$$

where $\phi(I / K) \cdot I$ represents the adjustment costs (in terms of lost output). Adjustment costs imply that capital is imperfectly mobile across sectors. This is important because it allows the model to capture how policies differently affect capital productivity and profits in various industries. Perfect mobility would imply that in response to a new policy, capital instantly is reallocated so as to equate the marginal products of capital and cause similar profit impacts throughout the economy. The adjustment costs in equation (3.2) are given by the product $\phi(I / K) \cdot I$, where ϕ is a convex function of the rate of investment:

$$\phi(I^m/K^m) = \frac{(\xi^m/2)(I^m/K^m - \delta^m)^2}{I^m/K^m}, \quad m \in (s,e) \qquad (3.3)$$

and where δ^m is the rate of economic depreciation for capital stock m and ξ^m is the marginal cost of adjustment. The law of motion for each capital stock m is given by

$$K_{t+1}^m = I_t^m + (1 - \delta^m) K_t^m \qquad (3.4)$$

Profits and the Behavior of Firms

As indicated earlier, the value of the firm is the present value of after-tax profits net of new share issues. Capital income (before corporate taxes), denoted by π^b, is the value of net output less payments to labor and payments to energy and material inputs:

$$\pi^b = p^n Y - (1 + \tau^P) wL - p^E E - p^M M \qquad (3.5)$$

where $p^n Y$ is the value of net output (with p^n denoting the per-unit price of output net of output taxes), $(1 + \tau^P) wL$ is payments to labor, and p^E and p^M denote the unit prices of the energy and material composites. Payments for intermediate energy and material inputs reflect the prices and quantities of individual energy and materials inputs. Thus, for each industry j,

$$p_j^E E_j = \sum_i^{N_e} p_{ij}^E E_{ij}$$

where p_{ij}^E are the prices, net of intermediate input taxes, of individual energy inputs i to industry j, and E_{ij} are the quantities of these inputs to this industry. A similar equation holds for the individual material inputs.

The climate policies considered in this text directly affect input costs. For example, a carbon tax imposed on suppliers of fossil fuels will raise the prices of those fuels and thereby increase input prices p_{ij}^E for all industries j that use those fuels $i \in$ (coal, oil, gas). (We describe this further in chapter 5.) Firms that do not directly use fossil fuels will also be affected to the extent that they use carbon-intensive inputs, as the prices of those inputs will also rise in response to emissions pricing.[7] This applies, for example, to electric utilities and to firms in the refining industry. Other things being equal, increases in $p^E E$ in a given industry will reduce capital income π^b in that industry. In response to these policy-induced changes in input prices, cost-minimizing producers will generally alter the mix of inputs to production.

After-tax profits π are equal to after-tax capital income less interest payments on outstanding debt and property tax payments plus tax deductions.[8] Under some of the policies considered, firms receive lump-sum transfers. Under some of the designs for a carbon tax, for example, firms receive lump-sum payments to offset their carbon tax obligations. Similarly, under some designs for a cap-and-trade program, firms receive some free allowances; firms getting free allowances are in effect receiving a lump-sum payment equal to the value of these allowances. These payments represent lump-sum increments to after-tax profits.

A cash-flow identity links the sources and uses of revenues by firms:

$$\pi_t + (\text{DEBT}_{t+1} - \text{DEBT}_t) = \text{DIV}_t + \text{IEXP}_t + \text{SR}_t \tag{3.6}$$

The firm receives cash from profits and new debt issue, $\text{DEBT}_{t+1} - \text{DEBT}_t$. It uses its cash to finance dividend payments DIV_t, investment expenditures IEXP_t, and share repurchases SR_t. See appendix B for derivations of each expression in equation (3.6). Climate policies can influence all the right-hand-side elements of the earlier equations. In addition to influencing investment as indicated earlier, they affect dividends, which are modeled as a constant fraction of after-tax profits. They also influence share repurchases, as these must make up the difference between sources of revenue (the left-hand side of (3.6)) and the uses of revenue (the right-hand side of the equation).

Households (stockholders) require firms to offer them a rate of return comparable to the rate of return on owning private or public debt.[9] The after-tax return to stockholders is equal to the after-tax value of dividends and capital gains (net of share repurchases); this must be equal to the after-tax return of an investment of the same value, where the investment has a rate of return equal to the return on bonds:

$$(1-\tau^e)\text{DIV}_t + (1-\tau^v)(V_{t+1} - V_t + \text{SR}_t) = (1-\tau^b)rV_t \tag{3.7}$$

where τ^e, τ^v, and τ^b represent personal tax rates on dividend income, capital gains, and interest income, respectively. An expression for the value of the firm can be derived from equation (3.7) such that the value of the firm is equal to the discounted sum of after-tax dividends and share repurchases.

Features Specific to the Oil Industry

The model captures some of the unique features of the crude oil industry. One of these is the non-renewable nature of the oil supply. Production in this industry involves stock effects: as reserves are depleted over time, producers need to introduce more costly extraction methods in existing fields (e.g., introduce secondary recovery techniques to replace primary recovery extraction methods) or move to new fields that are less fecund.[10] In either case, the depletion of reserves is accompanied by an increase in unit costs.[11] In making current decisions about rates of extraction, forward-looking producers in these industries take into account the impact of their actions on future costs.[12]

Let γ_0 represent the total factor productivity of oil producers in the benchmark year. Productivity in any given period t is given by

$$\gamma_t = \gamma_0[1 - (Q_t / \bar{Q})^\varepsilon] \tag{3.8}$$

where Q_t represents cumulative extraction at the beginning of period t, $\bar{Q}$ is the estimate of the total stock of recoverable reserves in the benchmark year, and ε is a curvature parameter that captures the rate of decline in productivity. Cumulative extraction follows the equation of motion: $Q_{t+1} = Q_t + Y_t$.

A second feature is the inclusion of a backstop substitute for crude oil. The backstop is modeled as a perfect substitute for oil. The backstop is not economically competitive with crude oil initially, but eventually becomes so as the costs of supplying oil increase. The assumptions about the backstop fuel in our reference case are described in chapter 4.

A third feature relates to the treatment of the price of oil. For all goods and services other than oil, the output prices are endogenous, reflecting the scarcity of factors of production (labor and capital) and the associated costs of supply. Oil is an exception: the model treats the real price of oil as exogenous.[13] In the model, international oil markets are fully integrated.[14] Accordingly, imported oil is treated as a perfect substitute for domestically supplied oil, and the domestic price is equal to the exogenously specified world price.[15]

Carbon Dioxide Emissions

The model calculates CO_2 emissions Z_t based on the carbon content of fossil fuels employed as inputs in production. Emissions are the product of CO_2 emissions coefficients and the quantity of fuels purchased. By attributing emissions to the carbon content of the three fossil fuels (oil, coal, and natural gas), the model accounts for both the emissions that result when the primary fuels (e.g., crude oil) are combusted and the emissions stemming from combustion of their downstream products (e.g., diesel fuel or gasoline).[16]

CONSUMER BEHAVIOR

Our modeling of consumer behavior captures key aspects of consumer choice: the choice between work hours and leisure time, the choice between current consumption and saving for future consumption, and the allocation of current consumption expenditure across various consumer goods and services.

The model specifies a single, representative household to capture consumer behavior. This specification enables us to assess average impacts, but it does not convey how impacts differ across different types of consumers. In chapter 7 we explore the different impacts across household groups by linking the E3 model with a model that captures differences across households in their expenditure patterns and sources of income. With the linked models we examine the different impacts of climate policies across household groups as well as the potential to avoid uneven impacts through different forms of compensation such as personal income tax cuts or lump-sum rebates.

Utility and Consumption

The household has constant relative risk aversion (CRRA) utility over *full consumption* C_s in each period s.[17] Expected lifetime utility in period t is a function of this consumption over time:

$$U_t = \sum_{s=t}^{\infty} \beta^{s-t} \frac{1}{1-\sigma} C_s^{1-\sigma}$$

(3.9)

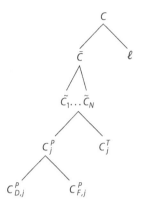

FIGURE 3.3

E3 Consumption Structure

where β represents the preference for consumption across time and σ is the coefficient of relative risk aversion. With this functional form, σ also represents the inverse of the intertemporal elasticity of substitution in consumption, and thus it affects the substitutability of full consumption across time. Full consumption is a composite of consumption of goods and services and of leisure, according to the nested structure indicated in figure 3.3. At the top level, full consumption is a composite of the consumption-good composite $\bar{C}$ and leisure ℓ:

$$C = \left[\bar{C}^{\frac{\eta-1}{\eta}} + \alpha_\ell^{\frac{1}{\eta}} \ell^{\frac{\eta-1}{\eta}} \right]^{\frac{\eta}{\eta-1}} \tag{3.10}$$

The parameter η represents the elasticity of substitution between consumption and leisure, and α_ℓ is a leisure intensity parameter.

Here $\bar{C}$ is a composite of N_c consumer goods $\tilde{C}_j (j=1,...,N_c)$. The functional form of the $\bar{C}$ composite is Cobb Douglas; the household chooses total spending on each good j to minimize total costs, given this functional form. Table 3.2 lists the 24 consumption goods and the benchmark level of spending on each good. Spending on each good involves payments to producers of the good involved as well as expenditure on associated transportation and trade services, as described later.

TABLE 3.2 Personal Consumption Expenditures by E3 Category, 2013

CONSUMPTION CATEGORY	CONSUMPTION[a]	PCT OF TOTAL CONSUMPTION
Motor vehicles	549.0	4.8
Furnishings and household equipment	394.5	3.4
Recreation	1,022.1	8.9
Clothing	425.8	3.7
Health care	2,372.1	20.7
Education	277.1	2.4
Communication	283.1	2.5
Food	750.3	6.5
Alcohol	124.7	1.1
Motor vehicle fuels (and lubricants and fluids)	381.8	3.3
Fuel oil and other fuels	26.6	0.2
Personal care	245.3	2.1
Tobacco	108.0	0.9
Housing	1,780.9	15.5
Water and waste	136.4	1.2
Electricity	169.1	1.5
Natural gas	51.2	0.4
Public ground	42.3	0.4
Air transportation	49.5	0.4
Water transportation	3.2	0.0
Food services and accommodations	714.7	6.2
Financial services and insurance	826.7	7.2
Other services	700.5	6.1
Net foreign travel	44.2	0.4
Total	**11,478.9**	**100.0**

[a] In billions of 2013 dollars.

At the lowest level of the nest, households choose whether to purchase goods (or services) from domestic or foreign suppliers. Total consumption for good j, denoted by C_j^P, is a constant-elasticity-of-substitution composite of domestic and foreign-produced goods $C_j^P = c_j(C_{D,j}^P, C_{F,j}^P)$. Armington trade elasticities determine how flexible the consumer is in choosing between domestic and foreign goods when relative prices change.[18]

Prices and the Budget Constraint

When consumers purchase consumption good j, the price paid reflects transportation costs and the wholesale and retail markups charged by vendors and retailers. Therefore, the revenue from a dollar spent on a consumption good by the household will accrue to several parties: the producer(s) of the good, the transportation sector(s) used to deliver the good to market, and the wholesale and retail vendors that sell the good to the household. Trade and transportation margins are assumed to be constant at the benchmark year level; α_j^T represents the share of trade and transport margins for each good j. Some goods, such as manufactured goods, will have large transportation and trade margins, whereas other goods such as consumer services will have zero transportation and trade margins. A Leontief aggregation function combines consumption and transportation and trade costs; thus, the unit price, exclusive of taxes, of good j, $\hat{p}_j$, is a weighted average of the producer price p_j^P and the trade and transportation costs p_j^T:[19]

$$\hat{p}_j = (1-\alpha_j^T)p_j^P + \alpha_j^T p_j^T \tag{3.11}$$

Consumption spending on goods and services is subject to ad valorem and/or fixed excise taxes. For some goods and services, the effective price to the consumer will also reflect tax deductions or credits associated with purchases of the good in question.[20] The after-tax/subsidy price $\tilde{p}$ is a function of the ad valorem tax τ^{ca}, the excise tax τ^{ce}, the income tax deduction s^d, the average personal income tax rate that applies to deductions τ^s, and the level of tax credits s^c:[21]

$$\tilde{p}_j = (1-\tau^s s_j^d)((1+\tau_j^{ca})\hat{p}_j + \tau_j^{ce}) - s_j^c \tag{3.12}$$

Given Cobb-Douglas consumption shares α_j^C, the after-tax composite price of consumption $\bar{p}$ is

$$\bar{p} = \prod_{j=1}^{N_c} \tilde{p}_j^{\alpha_j^C} \tag{3.13}$$

The household's budget constraint is

$$W_{t+1} - W_t = \bar{r}_t W_t + (1 - \tau^L) w_t l_t + G_t^T + G_t^C - T_t^L - \bar{p}_t \bar{C}_t \tag{3.14}$$

where W represents financial wealth (the value of equity in firms and the value of private and public bonds) and $\bar{r}$ is the after-tax return on financial wealth. The term $\bar{r}_t W_t$ therefore represents the after-tax capital income of the household. After-tax labor income is given by the term $(1 - \tau^L) w_t l_t$ where τ^L is the tax rate on wage income (inclusive of employee payroll taxes), w_t is the nominal pretax wage rate, and l_t represents aggregate labor supply. Households are endowed with time $\bar{l}$, which is allocated to either leisure or labor: $\bar{l} = l_t + \ell_t$. Households receive fixed transfers from the government with a value of G_t^T and in some policy cases receive a lump-sum carbon tax rebate of G_t^C. Finally, lump-sum taxes T_t^L constitute a portion of the household's tax obligations in each period. The price of consumption is $\bar{p}_t$; hence $\bar{p}_t \bar{C}_t$ is the value of total consumption. The left-hand side of the equation is household saving, the increment to financial wealth from period t to period $t+1$.[22]

GOVERNMENT

Federal, state, and local governments are modeled as a single agent. The government levies taxes and issues debt (owned by households). Revenues are used to finance government purchases of goods and services (including capital goods), government purchases of labor, and transfers to households.

Government Revenues

Government revenues come from tax receipts and deficit financing. Real government debt is assumed to grow exogenously at the steady-state growth rate. The nominal debt is equal to the real level of debt times the price level, and the level of deficit financing in any given period is the difference between the next period's nominal debt and the current level of nominal debt.

Tax revenues consist of tax receipts from households and firms. Households pay wage income taxes, employee payroll taxes, capital income taxes, and lump-sum taxes, as well as sales taxes on consumption goods and services (including gasoline). Firms pay corporate income taxes, employer payroll taxes, property taxes, and input taxes, as well as tariffs on imported inputs and goods. When introduced, the carbon tax represents another tax paid by firms. Tax receipts are offset by tax expenditures in the form of subsidies on tax-favored consumption goods or services.

Government Expenditures

The government employs capital, labor, and intermediate inputs to produce publicly provided goods and services. In the model, these goods and services do not enter the household's utility function, although of course they affect well-being in reality. To control for possible effects of these goods and services on utility, the model holds the value of these services fixed in real terms across policy experiments. In general, government purchases of intermediate inputs, the government capital stocks, and the government's use of labor are all assumed to grow exogenously at the steady-state growth rate, although we do allow for some substitution across intermediate inputs and labor to minimize the cost of providing the fixed level of publicly provided goods and services. Figure 3.4 displays the government production function. See appendix B for details on the modeling of government expenditures.

In addition to the government's investment expenditures (to replace depreciated capital), its purchases of intermediate inputs, and its direct expenditures on labor inputs, the government provides transfers to the household, as described in the consumer behavior section of appendix B. The real level of government transfers is held fixed across policies (growing at the steady-state growth rate). The government also makes interest payments to the household on the levels of public debt owned by the household. Finally, among the climate change policies considered, some involve the government's rebating additional revenues to the household through lump-sum payments (G_t^C in the household budget constraint earlier).

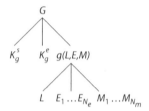

FIGURE 3.4

E3 Government Production Structure

The Government Budget Constraint and Revenue Neutrality

The government faces a budget constraint that requires total government expenditures to equal total government revenues plus new debt issue.

Climate policies can generate gross revenues in the form of carbon tax payments, receipts from auctioned emissions allowances, or revenues associated with increased gasoline taxes. At the same time, climate policies can negatively influence the revenues generated by other, existing taxes insofar as they reduce capital and labor incomes. This is the *tax-base effect* discussed in chapter 2. The government's budget constraint needs to account for the *net* revenue generated by climate policies, that is, the gross revenue minus any loss of revenue from existing taxes.

In principle, the government's budget constraint could be satisfied through adjustments in government expenditure on goods and services, government transfers, debt issue, or tax revenues. However, in the policy cases we consider, we maintain the same paths of real government spending, real transfers, and new debt issue as those in the reference case. Since these components of the budget constraint do not change from their values in the reference case, tax revenue also must not change if the government's budget constraint is to be satisfied. In our policy experiments, government revenues are recycled to keep tax revenues at reference case levels and thereby ensure that the budget constraint is met. The policies are therefore revenue-neutral: any potential net revenue from a given policy is offset through revenue recycling in the form of cuts in the tax rates on labor or capital income, lump-sum rebates offered to the households, or a combination of the two.

In each case, tax rate cuts or lump-sum rebates are modeled as permanent changes of a given amount rather than cuts and/or rebates that vary period by period. When tax cuts are employed, the cuts are fixed through time. Similarly, when lump-sum rebates are used, the size of the rebates are specified as increasing at a constant rate over time. The tax cuts and/or rebates are scaled so that the present value of the recycled revenue is equal to the present value of the net carbon revenue.[23]

THE FOREIGN ECONOMY

In several respects, the modeling of the rest of the world parallels that for the U.S. economy. As with the U.S. economy, the foreign economy has a representative household and government. The structural details of the foreign household and government mirror those for the U.S.

The modeling of foreign production is also similar to that for the U.S., although foreign production is more highly aggregated. Foreign production is carried out by a representative firm, as opposed to thirty-five distinct firms. This firm produces a range of intermediate inputs, consumer goods, and capital goods—the same categories of goods as those manufactured in the U.S. These goods are both consumed in the foreign economy and exported to the U.S. The supply of oil by foreigners on the world market is assumed to be infinitely elastic: that is, the foreign producer supplies oil to the world market at an exogenously specified world price. Appendix B describes in detail the general foreign production sector and the foreign oil production sector.

Import and export prices are converted to or from the foreign currency with the exchange rate e. This exchange rate adjusts to assure balanced trade in each period: the total value of exports to the foreign economy equals the total value of imports from the foreign economy.

By influencing domestic supplies and demands, U.S. policies affect the equilibrium prices and outputs of the goods produced abroad. In the policy simulations considered in this book, it is assumed that the changes in policy take place in the U.S. only: the foreign-economy's policy environment is assumed to remain unchanged. In the model, the impacts of climate policy on international trade and competitiveness depend on policy design. We concentrate on designs that maintain a level playing field for domestically produced and imported fuels.

EQUILIBRIUM

Equilibrium in the E3 model is attained when supply and demand in all markets is attained in all periods and the intertemporal optimization conditions for the household and firms are satisfied. The requirements of equilibrium in each period are:

1. For all commodities i, total supply of domestic commodity i equals total domestic demand of the commodity for consumption, investment, intermediate inputs, plus foreign demand (exports).
2. Total labor demanded by firms and government equals total labor supply.
3. Total private savings is equal to the sum of new debt issue (public and private) and the change in the equity value of firms.
4. The value of exports equals the value of imports.

These conditions must also hold for the foreign economy.

The conditions are met through adjustments in each period in prices, the nominal wage, the interest rate, and the exchange rate. Because these conditions are interdependent—the interdependence being the essence of general equilibrium—the equilibrating values of prices, the wage rate, the interest rate, and the exchange rate must be solved for simultaneously.

In fact, the interdependence applies across time as well as across markets in any given period. As mentioned earlier, the model's agents are forward-looking, taking expected future prices into account when they make current decisions. Hence, current decisions depend not only on current market (supply and demand) conditions but also on future market conditions. This implies that all the equilibrium prices are connected: the equilibrium prices in any given period depend on the equilibrium prices of all other periods, and prices are linked through the intertemporal first-order conditions for the households and firms. Equilibrium in the E3 model can be attained only when these intertemporal conditions hold.

See appendix B for a complete description of the intertemporal conditions for households and firms.

The solution technique employed to solve for the equilibrium prices solves simultaneously for all the prices in all periods, including the steady state, so that markets clear in all periods and the intertemporal conditions of the model hold. The model is formulated as a mixed complementarity problem and solved in GAMS using the PATH solver.[24] The mixed complementary problem organizes each of the excess demand conditions into a set of inequality constraints. The general equilibrium results when a set of prices is found that satisfies all the constraints.

When solving the model, we jointly solve for the steady state and transition path from the benchmark model year. The steady state is used to derive terminal terms for the transition path. We use 151 periods to ensure convergence of the economic variables on the transition path to the new steady state. The equilibrium solution for a given policy simulation is usually obtained in approximately one hour.

4

DATA, PARAMETERS, AND THE REFERENCE CASE PATH

To assess the impacts of potential U.S. climate-change policies in the E3 model, we adopt a multistep process, one that begins with the collection of primary data and parameters and ends with comparisons of the outcomes under various policy initiatives with the outcomes in the reference (business-as-usual) case.

Figure 4.1 displays these steps. We start by collecting raw data on all the variables in the model for a single benchmark year, 2013. This includes each industry's intermediate inputs, factor inputs and payments, and taxes paid. The initial data also include the components of final demand—consumption, investment spending, government expenditures by industry, and consumption by personal consumption expenditure (PCE) category. In the second step, we convert the raw data to a consistent social accounting matrix (SAM), as described later. In a third step, we combine the newly created SAM with parameters of the production and utility functions. Some of the parameters are taken from existing empirical studies, while the remaining parameters are identified by a "replication restriction"—the requirement that, under business-as-usual conditions, the behavior of the model's agents match the observed behavior.

At this point we have a complete dataset. When it is used as input for the model, the output from the model matches the benchmark information on supplies and demands of factors and goods.

When the benchmark data are used as input, the model generates a *base case time-path* of prices and output. This time-path exhibits balanced growth: all quantities grow at a single exogenously specified rate,

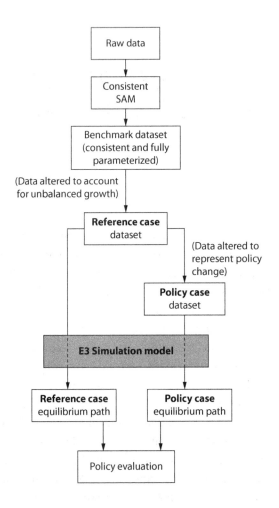

FIGURE 4.1

Links from Raw Data to Policy Evaluation

and relative prices do not change. However, in the base case the time paths for many important variables, such as CO_2 emissions and oil prices, do not match forecasts for the next several decades—for example, forecasts by the Energy Information Administration's (EIA's) *Annual Energy Outlook* (AEO) (Energy Information Administration, 2016). Hence the base case does not form the most realistic path against which we can compare the outcomes under policy changes. To create a more realistic comparison path, we introduce a few changes to the model or its parameters. We refer to the resulting time path as the *reference case time-path*.

Finally, as described in later chapters, we alter some inputs to the model or introduce some constraints to simulate a range of climate policies. For example, we place a price on carbon via a carbon tax. After performing simulations of the climate policies, we compare the outcomes under these policies with those arising in the reference case.

FROM PRIMARY DATA TO A SOCIAL ACCOUNTING MATRIX

Elements of the Social Accounting Matrix

A SAM serves as the numerical backbone for most CGE models. With its double-entry bookkeeping structure, the SAM provides a snapshot of income and expenditure flows for each industry in the model. The flows link final demand, intermediate input transactions, and payments to factors (value added).

Figure 4.2 displays the SAM structure for the E3 model. The matrix $\bar{X}_D$ is an input-output table, an $N \times N$ matrix indicating the value of purchases of each domestic commodity i by each industry j ($i, j = 1, \ldots, N$). The similar matrix $\bar{X}_F$ indicates the value of purchases of each foreign-produced commodity i by each domestic industry j. Summing the columns of $\bar{X}_D$ and $\bar{X}_F$ yields total intermediate inputs of each commodity by industry. In energy and environmental policy modeling, this information is crucial, as it can be used to calculate total energy inputs to each industry.

The $N \times 1$ column vectors $\bar{C}_D$, $\bar{I}_D$, $\bar{G}_D$, $\bar{G}_D^I$, and $\overline{EX}$ represent the demand for each domestic commodity for the purposes of consumption, investment, government intermediate inputs, government investment, and exports.

The $1 \times N$ row vectors $\bar{L}$ and $\bar{K}$ represent the payments to households for the use of labor and capital, respectively.[1] The $1 \times N$ row vector $\bar{T}$ represents input and output taxes by industry (sales taxes, payroll taxes, intermediate input taxes).

The SAM indicates flows of income and expenditure across industries, and between industries, households, and the government. The elements

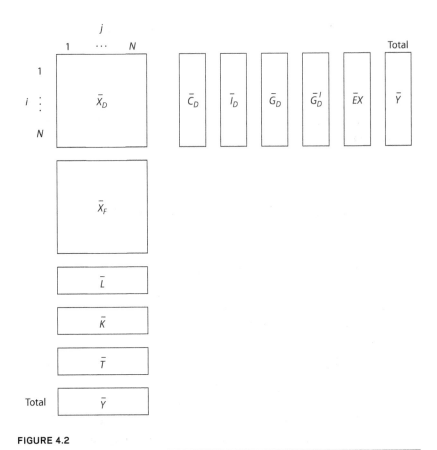

FIGURE 4.2

Social Accounting Matrix Structure

are expressed in values, rather than in typical units of quantities (such as barrels of oil or short tons of coal). For any given good, one can compute the quantity in terms of a desired quantity unit by dividing the value in the SAM by the unit price of the good in the benchmark year.

For the E3 model, for each good and service, we define the quantity unit as the quantity that has a value of $1 in the benchmark. Thus, a benchmark expenditure of X dollars on a given good corresponds to expenditure on X units of the good. Where we provide numbers later, for convenience we often display the numbers in billions, so that a displayed value or quantity of N denotes N billion dollars or N billion quantity units.

Building the Initial SAM from Primary Data

Our first step in constructing the SAM is to obtain industry-level transaction data from the Bureau of Economic Analysis (BEA). The BEA annually publishes the Input-Output Accounts as part of its Industry Economic Accounts. These accounts primarily comprise a Use matrix and a Make matrix.[2] The Use matrix describes, for seventy-one commodities in the annual tables, the value of total (domestically and foreign supplied) commodity use by industry as well as the total value of (domestically and foreign supplied) commodity use by each final good demand category.[3] To disaggregate the Use matrices into domestically supplied and foreign-supplied commodities, we use the supplemental Import Use matrix, which mirrors the Use matrix except that it only shows the use of imported commodities. By subtracting the two matrices, we obtain a Domestic Use matrix with final good demand vectors $\bar{C}_D$, $\bar{I}_D$, $\bar{G}_D$, $\bar{G}_D^I$, and $\overline{EX}$.

The Make matrix describes, for seventy-one commodities, how much of each commodity is "made" by each industry. While most industries make one primary commodity, some industries, especially manufacturing industries, make many commodities. For example, the plastics and rubber industry makes goods that are classified as sixteen distinct commodities (out of a total of seventy-one). To obtain the input-output matrices $\bar{X}_D$ and $\bar{X}_F$, which are commodity-by-commodity matrices, we convert the Make matrix into a Percentage Make matrix, which expresses the percentage of each industry's output in the form of each commodity. We then multiply the domestic and foreign Use matrices by the Percentage Make matrix to obtain the input-output matrices $\bar{X}_D$ and $\bar{X}_F$.[4]

To get the full SAM, we need to append to these input-output matrices and final good demand vectors the data on value added, consisting of payments to factors (factor incomes) and direct factor taxes. The factor income vectors $\bar{L}$ and $\bar{K}$ are constructed from several sources. The primary source for the labor income vector is the Use matrix, which includes total labor compensation by industry. Our SAM vector $\bar{L}$ adjusts for employer payroll taxes.

To construct capital income, we first obtain capital stocks and depreciation rates from the BEA's Fixed Asset Accounts.[5] The E3 model assigns capital to two categories: (a) structures and (b) equipment/intellectual property products. Given the depreciation rates, along with marginal tax rates on capital income, tax depreciation rates, debt ratios, payout ratios, and interest rates, we construct industry-specific, depreciation-adjusted, tax-adjusted returns on each type of capital. Multiplying this capital return by the capital stock and summing over the types of capital yield the vector $\bar{K}$ in the SAM.

To construct the return to capital used to create the SAM vector $\bar{K}$, we need information on marginal tax rates, tax depreciation rates, debt ratios, payout ratios, and interest rates, along with capital stock levels and capital depreciation rates. Marginal tax rates on wages, dividends, capital gains, and interest income derive from the NBER-TAXSIM database.[6] Taxes on corporate income are set to 40 percent (equal to the federal rate of 35 percent plus an average state corporate income tax of 5 percent). Property taxes are derived by dividing the total property taxes collected in 2013 by the total value of the capital stock in 2013, which implies that all types of capital face the same property tax rate.[7] Payroll tax rates are also assumed to be constant across industries: employer- and employee-paid payroll taxes are divided by labor income to determine marginal employer and employee payroll taxes.[8] Table 4.1 lists the marginal tax rates in the model.

Tax depreciation rates δ_D^s and δ_D^e are derived from the tax code as follows. Using the General Depreciation System (GDS) guidelines, we set the tax depreciation rates so that structures depreciate in 39 years and

TABLE 4.1 E3 Marginal Tax Rates (in percent)

Individual Income Taxes	
Labor	26.78
Dividends	27.73
Interest	31.78
Capital gains	26.12
Payroll Taxes	
Employer	6.32
Employee	6.95
Firm Taxes	
Corporate income	40.00
Property taxes	1.09

equipment/intellectual property depreciates in 7 years.[9] For each industry, the debt ratio b is the fraction of the value of the industry's capital stock that is held as debt, and the payout ratio a is the fraction of after-tax profits paid to households as dividends. We use information on debt and payout ratios by sector provided by Damodaran Online.[10] Table 4.2 displays tax depreciation rates, debt ratios, and payout ratios by industry.

The tax vector $\bar{T}$ is constructed by including employer payroll taxes and total intermediate input taxes by industry.[11]

Producing a Consistent SAM

The initial SAM constructed from the various data sources fails to satisfy several general equilibrium requirements of the model in the benchmark year: market clearing, zero economic profits, income-expenditure balance, and trade balance. *Market clearing* requires that total demand for industry i's output—the sum across the ith row of the SAM—equal the output of the ith industry. The *zero-profit condition* is satisfied when, for each industry, total revenue—the sum across a given industry row—equals total expenditure—the sum down the column for that industry. *Income-expenditure balance* is achieved when total factor income equals total final good demand, that is, when the budget constraints of the representative agent and the government are satisfied.[12] *Trade balance* requires the value of exports to equal that of imports. To generate a SAM that meets these conditions, we make the following adjustments to the data:

- Adjust rows and columns of the domestic input-output table $\bar{X}_D$ so that the zero-profit condition holds for every industry.
- Scale consumption to make it consistent with the household budget constraint and the income-expenditure balance condition.
- Scale private fixed investment spending so that the rate of investment in each industry is consistent with balanced growth.
- Scale lump-sum tax payments from households so that the government budget constraint holds.
- Scale exports to be consistent with zero trade balance.

For more details on the consistency procedure, please refer to appendix C.

TABLE 4.2 Depreciation Rates, Tax Depreciation Rates, Debt Ratios, and Dividend Ratios by E3 Industry

INDUSTRY	DEPRECIATION RATES		TAX DEPRECIATION RATES		DEBT RATIO	DIVIDEND RATIO
	δ^s	δ^e	δ_D^s	δ_D^e	b	a
Oil extraction	0.076	0.161	0.026	0.143	0.192	0.334
Natural gas extraction	0.076	0.161	0.026	0.143	0.192	0.334
Coal mining	0.042	0.161	0.026	0.143	0.750	0.750
Electric transmission and distribution	0.022	0.083	0.022	0.083	0.615	0.676
Coal-fired electricity generation	0.022	0.083	0.022	0.083	0.615	0.676
Other-fossil electricity generation	0.022	0.083	0.022	0.083	0.615	0.676
Nonfossil electricity generation	0.022	0.083	0.022	0.083	0.615	0.676
Natural gas distribution	0.023	0.083	0.023	0.083	0.615	0.676
Petroleum refining	0.032	0.100	0.026	0.100	0.192	0.300
Pipeline transportation	0.028	0.147	0.026	0.143	0.453	0.750
Mining support activities	0.064	0.139	0.026	0.139	0.192	0.300
Other mining	0.042	0.161	0.026	0.143	0.350	0.326
Farms, forestry, fishing	0.025	0.145	0.025	0.143	0.000	0.000
Water utilities	0.023	0.083	0.023	0.083	0.487	0.540
Construction	0.026	0.177	0.026	0.143	0.237	0.198
Wood products	0.033	0.154	0.026	0.143	0.271	0.750
Nonmetallic mineral products	0.031	0.125	0.026	0.125	0.271	0.750
Primary metals	0.032	0.085	0.026	0.085	0.489	0.629
Fabricated metal products	0.032	0.111	0.026	0.111	0.455	0.719
Machinery and misc. manufacturing	0.031	0.214	0.026	0.143	0.146	0.356
Motor vehicles	0.031	0.221	0.026	0.143	0.181	0.399
Food and beverage	0.032	0.118	0.026	0.118	0.174	0.546
Textile, apparel, leather	0.030	0.125	0.026	0.125	0.073	0.243
Paper and printing	0.032	0.135	0.026	0.135	0.465	0.579
Chemicals, plastics, and rubber	0.031	0.123	0.026	0.123	0.208	0.326
Trade	0.026	0.201	0.026	0.143	0.154	0.286
Air transportation	0.024	0.078	0.024	0.078	0.369	0.184
Railroad transportation	0.022	0.109	0.022	0.109	0.179	0.337
Water transportation	0.032	0.089	0.026	0.089	0.425	0.276
Truck transportation	0.030	0.190	0.026	0.143	0.324	0.218
Transit and ground passenger transportation	0.023	0.149	0.023	0.143	0.205	0.361
Other transportation and warehousing	0.022	0.130	0.022	0.130	0.205	0.361
Communication and information	0.025	0.175	0.025	0.143	0.187	0.335
Services	0.021	0.205	0.021	0.143	0.750	0.336
Real estate and owner-occupied housing	0.024	0.171	0.000	0.000	0.000	0.000

PRIMARY PARAMETERS

The model's primary parameters are not identified by consistency requirements of the SAM. Rather, they are chosen based on empirical estimates from various studies. For a list of the primary parameters, see appendix C.

Exogenous Growth Rates and the Rate of Time Preference

In the model, technological progress takes the form of Harrod-neutral (labor-augmenting) technological change. We assume Harrod-neutral technological progress at an annual rate of 1.5 percent. This implies that, in the steady state, all quantity variables including GDP grow at 1.5 percent. The price level is specified as growing at 2 percent per year. We assume a real interest rate of 3 percent and a nominal interest rate of 5.06 percent.[13] Balanced growth in the base case implies that the discount factor β is a function of interest rates, exogenous growth rates, and the intertemporal elasticity of substitution $\beta = (1 + g_r)^\sigma (1+r)^{-1}$.

Production Parameters

The elasticities of substitution in production jointly determine the price elasticity of supply and the elasticity of demand for intermediate inputs, labor, and capital. The model derives these elasticities from estimates by Jorgenson and Wilcoxen (1996) and Jorgenson *et al.* (2013). We translate the estimates of parameters for translog cost functions into elasticities of substitution parameters to make them compatible with the constant-elasticity-of-substitution function form of our model.[14] The elasticities are reported in table 4.3.

In the absence of constraints or limitations in supply, the electricity produced from different generators would have the same price because they are perfect substitutes. However, regional variations in generating capacity by source, transmission constraints, ramping constraints, and storage limitations imply that electricity from different generators usually is not perfectly substitutable. Accordingly, we model the electricity from our three different generator types as imperfectly substitutable. We use an elasticity of substitution of 3, which is roughly consistent with annual

TABLE 4.3 Production Elasticities of Substitution

INDUSTRY	σ_e	σ_m	σ_{em}	σ_{lem}	σ_y	σ_k	σ_{df}
Oil extraction	0.95	0.81	0.7	0.7	0.5	1	∞
Natural gas extraction	0.95	0.81	0.7	0.7	0.5	1	1.10
Coal mining	0.51	1.03	0.7	0.7	0.5	1	1.10
Electric transmission and distribution	3.00	0.72	0.4	0.7	0.3	1	1.10
Coal-fired electricity generation	0.20	0.20	0.2	0.2	0.3	1	1.10
Other fossil electricity generation	0.20	0.20	0.2	0.2	0.3	1	1.10
Nonfossil electricity generation	0.20	0.20	0.2	0.2	0.3	1	1.10
Natural gas distribution	0.89	1.27	0.4	0.7	0.3	1	1.10
Petroleum refining	1.60	1.05	0.7	0.7	0.5	1	1.50
Pipeline transportation	0.41	0.95	0.7	0.7	0.5	1	1.10
Mining support activities	0.89	0.88	0.7	0.7	0.5	1	1.10
Other mining	0.81	1.11	0.7	0.7	0.5	1	1.10
Farms, forestry, fishing	0.59	0.34	0.4	0.7	0.3	1	1.10
Water utilities	0.55	0.91	0.7	0.7	0.5	1	1.10
Construction	0.74	0.63	0.7	0.7	0.5	1	1.10
Wood products	0.36	0.34	0.7	0.7	0.5	1	1.10
Nonmetallic mineral products	0.70	0.91	0.7	0.7	0.5	1	1.10
Primary metals	1.64	0.40	0.7	0.7	0.5	1	1.16
Fabricated metal products	0.65	1.71	0.7	0.7	0.5	1	0.88
Machinery and misc. manufacturing	0.95	0.83	0.7	0.7	0.5	1	3.00
Motor vehicles	0.50	0.95	0.7	0.7	0.5	1	3.00
Food and beverage	0.55	0.41	0.7	0.7	0.5	1	3.12
Textile, apparel, leather	0.62	0.38	0.7	0.7	0.5	1	3.60
Paper and printing	0.56	1.15	0.7	0.7	0.5	1	1.10
Chemicals, plastics, and rubber	0.54	0.20	0.7	0.7	0.5	1	1.50
Trade	0.38	0.95	0.7	0.7	0.5	1	1.10
Air transportation	0.41	0.95	0.7	0.7	0.5	1	1.10
Railroad transportation	0.41	0.95	0.7	0.7	0.5	1	1.10
Water transportation	0.41	0.95	0.7	0.7	0.5	1	1.10
Truck transportation	0.41	0.95	0.7	0.7	0.5	1	1.10
Transit and ground passenger transportation	0.41	0.95	0.7	0.7	0.5	1	1.10
Other transportation and warehousing	0.41	0.95	0.7	0.7	0.5	1	1.10
Communication and information	0.25	1.08	0.7	0.7	0.5	1	1.10
Services	0.63	0.95	0.7	0.7	0.5	1	1.10
Real estate and owner-occupied housing	0.77	0.95	0.7	0.7	0.5	1	1.10

σ_e: Elasticity of substitution between energy inputs. *Source*: Jorgenson et al. (2013).
σ_m: Elasticity of substitution between material inputs. *Source*: Jorgenson et al. (2013).
σ_{em}: Elasticity of substitution between energy and material inputs. *Source*: Jorgenson and Wilcoxen (1996).
σ_{lem}: Elasticity of substitution between labor and intermediate inputs. *Source*: Jorgenson and Wilcoxen (1996).
σ_y: Elasticity of substitution between capital and labor-intermediate inputs. *Source*: Jorgenson and Wilcoxen (1996).
σ_k: Elasticity of substitution between different capital stocks.
σ_{df}: Elasticity of substitution between domestic and foreign intermediate inputs. *Source*: Feenstra et al. (2014).

substitution elasticities estimated from the detailed and disaggregated Haiku electricity model of Resources for the Future.

As mentioned earlier, the model incorporates capital adjustment costs—the costs of installing or removing capital. Earlier studies estimated large aggregate adjustment costs. See, for example, Summers (1981). More recent studies suggest that aggregate capital installation costs are small relative to other investment-related costs. See, for example, Hall (2004) and Cooper and Haltiwanger (2006). In our central case, we employ a compromise value of 7 for the adjustment cost parameters ξ_s and ξ_e.[15] We then test the sensitivity of our key results to both higher and lower adjustment costs.

Household Parameters

The elasticity of labor supply determines how much the household will alter its supply of labor (or change its leisure time) in response to any change in real wages. We set the elasticity of substitution between consumption and leisure η to yield a compensated elasticity of labor supply of 0.3 for our representative household in the benchmark data.[16] This value is at the high range of estimates for married men and single women (0.1–0.3) and in the middle of the range of estimates for married women (0.2–0.4).[17]

The representative household has constant relative risk aversion (CRRA) utility over full consumption. We use a value of 2 for the risk aversion parameter σ. This also implies an intertemporal elasticity of substitution in consumption of 0.5, a value between time-series estimates (Hall 1988) and cross-sectional studies (Lawrance 1991). To generate a value of leisure in the benchmark year, we assume that workers spend 66 percent of their time working. This fraction of time spent working implies a nonlabor income elasticity of labor supply of 0.25 and an uncompensated elasticity of labor supply of 0.05.

Armington Trade Elasticities

Armington trade elasticities determine the extent to which the household and firms will alter their demand between domestically produced and foreign-produced goods in response to changes in prices of these goods.

Feenstra et al. (2014) estimate elasticities between home and import goods for the U.S. and obtain evidence that the elasticity is greater than 1 for many industries.[18] For all of the industry goods and consumer goods estimated by Feenstra et al., we map the implied Armington elasticity to the E3 industries and consumer goods.[19] For foreign demand, we double the trade elasticities.[20] The values of the trade elasticities for industry goods are displayed in table 4.3. The values of the trade elasticities for consumer goods are translated from the production elasticities using an intensity IC matrix; this matrix (described in appendix B) indicates the intensities of producer goods used to produce each consumption good.

Consumption Taxes and Tax-Favored Consumption Subsidies

Taxes on consumption derive from the BEA's National Income and Product Accounts (NIPA) table 3.5: Taxes on Production and Imports. Federal, state, and local excise taxes are allocated to corresponding E3 consumption categories. Per-unit excise taxes are calculated by dividing total excise taxes by total consumption for each good. Table 4.4 reports excise taxes τ^{ce} as a percent of benchmark-year prices. Data on general sales taxes are also taken from NIPA table 3.5; sales taxes are assumed to be equal across goods (with the exception of motor vehicle fuels and housing). The rate of ad valorem consumption taxes τ^{ca} is equal to benchmark levels of sales tax revenue divided by total consumption (less consumption of exempt goods).

Tax-favored consumption subsidies derive from estimates by the Joint Committee on Taxation (2014) and the Internal Revenue Service (2015). These subsidies are mapped into the E3 consumer good categories.[21] Table 4.4 displays tax deductions s^d, the fraction of a unit of consumption that may be deducted from taxable income, and tax credits s^c, the reduction in the level of tax payments, expressed as a percent of benchmark-year prices.

Carbon Coefficients

As indicated in chapter 3, CO_2 emissions are represented as the product of CO_2 emissions coefficients and the levels of fossil fuel inputs in various

TABLE 4.4 Consumption Taxes and Subsidies by E3 Consumption Good

CONSUMPTION CATEGORY	AD VALOREM SALES TAX RATE (%)	EXCISE TAX[a]	TAX DEDUCTION[b]	TAX CREDIT[c]
Motor vehicles	4.2	1.8		0.1
Furnishings and household equipment	4.2			0.4
Recreation	4.2		6.1	
Clothing	4.2			
Health care	4.2	0.2	13.0	
Education	4.2		21.4	5.3
Communication	4.2			
Food	4.2			
Alcohol	4.2	12.1		
Motor vehicle fuels (and lubricants and fluids)		19.4	2.9	
Fuel oil and other fuels	4.2			
Personal care	4.2			
Tobacco	4.2	30.5		
Housing				
Water and waste	4.2	14.7		
Electricity	4.2	3.4		
Natural gas	4.2	5.6		
Public ground	4.2			
Air transportation	4.2	30.5		
Water transportation	4.2			
Food services and accommodations	4.2			
Financial services and insurance	4.2	2.2	26.0	0.2
Other services	4.2		97.5	0.5
Net foreign travel	4.2			

[a]Excise tax is expressed as percent of benchmark-year prices.
[b]Tax deduction equals fraction of consumption that is eligible for reductions in taxable income.
[c]Tax credit is expressed as percent of benchmark-year prices.

TABLE 4.5 U.S. Carbon Dioxide Emissions from Energy Consumption by Fuel and Sector, 2013

	RESIDENTIAL	COMMERCIAL	INDUSTRIAL	TRANSPORTATION	ELECTRIC POWER	TOTAL
Coal	0	4	142	0	1,571	1,717
Natural gas	277	178	462	49	444	1,410
Petroleum	67	40	347	1,756	23	2,233
Total	344	222	951	1,805	2,038	5,360

Source: U.S. Energy Information Administration, millions of metric tons.

industries. We obtain the coefficients using data on emissions from energy consumption by industry source in the benchmark year. Table 4.5 displays the distribution of emissions by fuel and source in 2013. To get the emissions factors for coal, we divide total emissions from coal in the electricity sector (from the EIA data) by the output of coal-fired generation in the E3 model; this gives the emissions factor from combustion of coal by coal-fired generators. Similarly, we divide the EIA's non-power-sector emissions from coal by the total output of the remaining sectors in the E3 model, to obtain a common coal-based emissions factor for these other industries.

To obtain the emissions factors pertaining to the combustion of natural gas, we divide EIA electricity-sector emissions from natural gas combustion by the output of the other-fossil generation industry in the E3 model; and we divide EIA residential- and commercial-sector natural gas emissions by the output of the E3 model's natural gas distribution sector (the sector that purchases natural gas to sell to households and non-industrial firms).

Finally, to obtain emission factors for natural gas combustion from industrial sources, and we relate the EIA data's industrial-sector natural gas emissions to the outputs of remaining sectors of the E3 model. We assign all oil emissions to the E3 petroleum refining sector, as this is the only sector that purchases crude oil in the E3 model. Table 4.6 displays the carbon coefficients by fuel and industry.

TABLE 4.6 Carbon Coefficients by Fuel and Industry

INDUSTRY	FUEL		
	COAL	NATURAL GAS	PETROLEUM
Oil extraction		0.022	
Natural gas extraction		0.022	
Coal mining		0.022	
Electric transmission and distribution			
Coal-fired electricity generation	0.070		
Other-fossil electricity generation		0.013	
Nonfossil electricity generation			
Natural gas distribution	0.024	0.009	
Petroleum refining	0.024		0.005
Pipeline transportation		0.022	
Mining support activities		0.022	
Farms, forestry, fishing			
Construction			
Air transportation		0.022	
Truck transportation		0.022	
Services	0.024	0.009	
Real estate and owner-occupied housing	0.024	0.009	
All other industries	0.024	0.022	

Note: Blanks indicate an industry does not directly use a given fuel.

THE REFERENCE CASE PATH

When we run the model with the consistent dataset just described, the model produces the base-case time-path. This time-path exhibits balanced growth: all quantities grow at the same constant rate, and relative prices remain unchanged.

As mentioned at the beginning of this chapter, the economy is not expected to exhibit balanced growth under business-as-usual conditions, and thus the base-case path is not the ideal time-path against which to compare the paths under policy changes. Accordingly, we introduce some changes to parameters and model structure to enable the model to generate a better time-path for the reference (business-as-usual) case.

Specifications for the Oil Industry

Three key changes are made to the model's treatment of the oil industry to enable it to generate the more realistic reference case path. First, the model is extended to address the nonrenewable nature of production of crude petroleum. With this change, the model accounts for stock effects: the increase in unit costs as a function of reduced reserves of these resources. The economic structure employed to account for stock effects is described in chapter 3. The key parameters that govern the magnitude of these effects are ε, the decline curvature parameters for each producer, and $\bar{Q}$, the benchmark-year level of recoverable reserves for oil. Using historical data, we estimate that the value of ε is 2. The total level of recoverable reserves of oil is set equal to 3 percent of current production.[22]

Second, for the reference case the model incorporates an exogenously increasing world price of crude oil. Based on forecasts from the EIA's Annual Energy Outlook (AEO) 2016, we assume that the real price increases annually by a constant amount equal to 1.24 percent of the benchmark price.

Third, the model incorporates a backstop technology for oil, as described in chapter 3. The technology is assumed to become commercially available in 2023. The backstop technology is assumed to be about 36 percent less productive than the benchmark productivity level of conventional oil production (chosen so that the long-run price of the backstop technology is 30 percent higher than the world oil price in the benchmark year). However, rising oil prices eventually make the backstop fuel competitive. We assume that the carbon content of a unit of the backstop fuel is equal to that of a unit of oil. As capital adjustment costs also apply to the backstop industry, the backstop fuel does not completely take over the market once its unit production costs are lower than the world oil price; rather, its market share expands gradually. Only in the very long run does the backstop fuel control the entire market.

Under central case values for parameters, the backstop fuel provides 46 percent of the total of domestic oil and backstop production by 2050. Imports of oil reach zero in 2072. In that year, domestic production of oil and the backstop fuel are sufficient to meet domestic demand. Prior to this point, the price of oil is determined by the exogenously specified world price of oil. Once this point is reached, the price of oil and its backstop are endogenous and are set to clear the market.

Fossil Fuel and Electricity Production

For the reference case, we also introduce some changes to parameters associated with the production of fossil fuels and electricity so that the prices of coal and natural gas will match the EIA's AEO 2016 time profiles. Specifically, we introduce a declining time profile for the scale parameter γ in the coal mining and natural gas industries.[23] The value of this parameter scales the total factor productivity of these industries, that is, the amount of output generated by given quantities of the inputs. In turn, total factor productivity affects unit costs and output prices. We choose the time profile of γ that causes the trend in output prices to match the trend in EIA prices.[24]

In the electric generation sectors, the AEO also predicts changes in natural gas generation and renewable generation. Regarding natural gas generation, we account for the fact that as efficient combined-cycle gas plants come on line and replace older plants, the average productivity of the gas generators is expected to increase. Through our choice of γ we cause total factor productivity in the other-fossil industry to match the AEO's predicted increases in output per unit of natural gas input.

With falling costs of renewables, the share of nonfossil generation is expected to increase significantly over the next 20 years. We increase the total factor productivity in this sector to match the predicted nonfossil generation shares in 2040.

Finally, to approximate the switch from coal to natural gas generation predicted by the AEO, we exogenously adjust the share parameter for the inputs of coal-fired and other-fossil electricity generation in the CES aggregation function for the electric transmission and distribution industry.

In the AEO projections, the consumption of carbon-intensive intermediate inputs and consumer products such as gasoline, electricity, and natural gas is forecast to grow at relatively low rates, implying that their shares of consumption will decline over time. Accordingly, we impose steadily declining share parameters. Specifically, for nonenergy industries, the share of energy goods in the intermediate good composite is reduced by 0.2 percent annually, and the share of energy goods in the household's consumption nest is lowered by 0.8 to 1.4 percent each year.

Reference Case Time Profiles

Figure 4.3 displays the reference case time profile of GDP and its consumption, investment, and government spending components over the years 2013–2050. Because of stock effects associated with extraction of crude petroleum, the steady increase in the world oil price, and the reference case assumptions on coal, natural gas, and electricity generation, the reference case path does not exhibit perfectly balanced growth. Nevertheless, the growth rates of GDP and its components are relatively constant through time.

Figure 4.4 shows the time profile for economywide emissions and electricity sector emissions in the reference case. Despite the projected

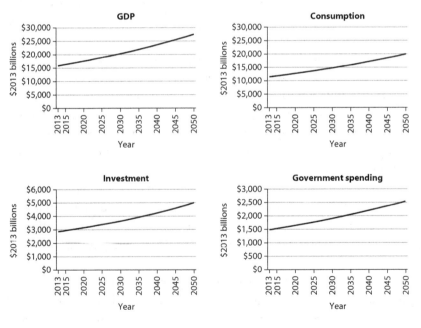

FIGURE 4.3

Reference Case GDP and Components, 2013–2050

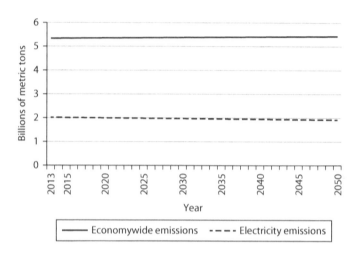

FIGURE 4.4

Reference Case Carbon Dioxide Emissions, 2013–2050

continued growth of the overall economy over the next few decades, econo-mywide emissions in the E3 model remain roughly constant over this time interval, reflecting the assumptions we imposed to match EIA pro-jections. Figure 4.5 displays how the shares of electricity coming from coal-fired, other-fossil, and nonfossil generators change over time. The coal-fired share declines as a result of the exogenous increases in other-fossil and nonfossil productivity and the assumed change in preferences between coal- and gas-fired electricity. These changing shares imply that emissions from the power sector decrease slightly between 2013 and 2050.

An important aspect of the reference case is the *marginal excess burden* (MEB) from various taxes in the model. The MEB of a given tax is the cost to the private sector from an incremental increase in that tax, when the revenue from this tax increase is returned to the private sector in a nondistortionary (lump-sum) fashion. The MEB gives an idea of the burden of the tax above and beyond the value of the revenue it raises. If, for example, the MEB of a given tax is $0.25, then if this tax were raised enough to generate $1 of revenue, the cost to the economy would be $0.25 after returning the dollar in lump-sum fashion.

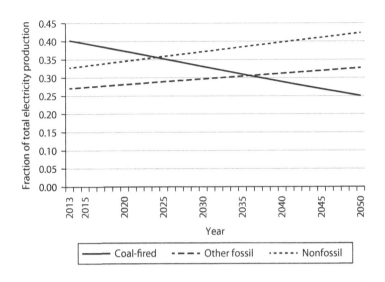

FIGURE 4.5

Reference Case Electricity Generation Shares, 2013–2050

The MEB values connect in important ways with the fiscal interaction issues presented in chapter 2. In particular, an economy in which taxes are highly distortionary (have high MEB values) will yield a large tax-interaction effect. Further, insofar as the MEBs of different types of taxes differ, the magnitude of the revenue-recycling effect will differ depending on the particular tax that is reduced. When revenues from an environmental tax or other revenue-raising environmental policy are used to finance cuts in a preexisting tax, the cost-reducing revenue-recycling effect will be larger, the greater the MEB value of the tax whose rate is cut.

Table 4.7 displays the MEB values that arise under central values for model parameters, for changes in payroll tax rates τ^p; wage income tax rates τ^l; individual capital income tax rates τ^b, τ^v, and τ^e; and corporate income tax rates τ^a. Payroll taxes have the lowest MEB value of $0.24. Wage income taxes have a slightly larger MEB value of $0.27. The difference is due to the fact that increases in taxes on wage income exacerbate preexisting distortions associated with tax credits and deductions. Capital taxes, both individual and corporate, have the highest MEB values: $1.04 and $1.08, respectively. The MEB on individual income taxes is a weighted

TABLE 4.7 **Marginal Excess Burden of Taxation**

TAX RATE	MEB
Payroll tax	$0.24
Individual income	$0.42
Wage income	$0.27
Capital income	$1.04
Corporate income	$1.08

average of the MEBs on wage income and individual capital income. These MEB values suggest that the overall economic costs from revenue-raising climate policies will be lowest in the case where the revenues are recycled in the form of cuts in the corporate income tax rate. We explore this issue numerically in later chapters.

Having conveyed the model's structure in chapter 3, and the model's data and reference case in this chapter, we now turn to the chapters that apply the model to assess alternative climate change policies.

III

POLICY APPROACHES AND OUTCOMES

—

5

TWO APPROACHES TO CARBON DIOXIDE EMISSIONS PRICING:
A CARBON TAX AND A CAP-AND-TRADE SYSTEM

Putting a price on carbon dioxide (CO_2) emissions addresses the climate-change-related external costs stemming from the combustion of carbon-based fuels. In doing so, emissions pricing aims to bring the prices of fossil fuels or of their refined products closer to the social cost of their use, where social cost includes both the usual market cost and the external cost. As indicated in chapter 2, appropriately scaled pricing brings net benefits to society: the gross benefits (the value of the environmental gains) exceed the gross costs (the economic sacrifices, ignoring the environmental impacts). In apparent recognition of this potential gain, an increasing number of politicians are calling for "putting a price on carbon." In an April 2016 article in *The New York Times*, World Bank President Jim Yong Kim stated that "putting a price on carbon is by far the most powerful and efficient way to reduce emissions." In that same article, International Monetary Fund Managing Director Christine Lagarde referred to carbon pricing as "the crown jewel" of mitigation efforts. The following October, Canada's Prime Minister Justin Trudeau stated that "carbon pricing is an effective way to reduce the pollution that threatens air quality and the quality of the ocean's water."

ELEMENTS AND IMPACTS OF EMISSIONS PRICING

Emissions pricing can be accomplished through a carbon tax or a cap-and-trade system. Each of these instruments allows for a range of designs in terms of stringency, the industries covered, the extent to which revenues are generated, and the way the revenues are dispersed. At the same time, *similarly designed* carbon tax and cap-and-trade policies generate similar economic incentives and thus can have very similar economic impacts.

A carbon tax generally is conceived as a tax on fossil fuels (oil, coal, and natural gas) or on refined fuel products, with the tax rate proportional to the carbon content of the fuel or product. In effect, a carbon tax is a tax on CO_2 emissions since the combustion of such fuels generates emissions proportional to the fuel's carbon content.[1]

Under cap and trade, the regulatory authority stipulates an allowable total quantity of emissions—the overall cap—for an interval of time extending into the future. In many policy proposals, the cap declines (i.e., gets tighter) over time. The regulator also introduces emissions allowances into circulation via free allocation or through an auction. Each allowance entitles the owner to a given quantity of emissions within a given period of time. The number of allowances introduced in each period is such as to authorize total emissions that match the specified overall cap for that period. Trading is a critical element of cap and trade; it promotes a single market price for all market participants.

The nature of uncertainty differs between a carbon tax and a cap-and-trade system. Under a carbon tax, the price of emissions is known—it is the carbon tax rate—while the extent of emissions reduction is determined by the market and thus is not known in advance. In contrast, under a cap-and-trade system, the quantity of emissions is known once the system is set up (assuming good compliance and enforcement); this is given by the number of allowances in circulation. In this case the price of emissions (the allowance price) is determined by the market and thus is uncertain. Groups that put greater emphasis on price certainty often favor a carbon tax over cap and trade, while organizations that stress the importance of certainty about emissions levels often prefer cap and trade.

Both of these instruments can be imposed at various points in the supply chain; that is, the *points of regulation* can be upstream, midstream, or downstream. An upstream carbon tax applies at the beginning of the carbon supply chain: at the mine mouth for coal and at the wellhead for oil and natural gas. A midstream tax would apply at the industrial user's gate. It could apply, for example, to petroleum refiners in proportion to the carbon content of the petroleum they purchase, to electric generators in proportion to the carbon content of their fuels, or to natural gas pipe-line companies in proportion to the carbon content of the natural gas they deliver. A downstream tax applies at the final point of combustion. Such a tax would apply to both large emitters such as coal-fired genera-tors and small emitters such as households that burn natural gas to heat their homes. Under cap and trade, the points of regulation can also be upstream, midstream, or downstream, depending on whether the policy requires allowances to be submitted by the firms that extract fossil fuels, the purchasers of fossil fuels, or the combustors of refined fuels.

No matter where the points of regulation might be, emissions pricing induces emissions-reducing substitutions all along the supply chain. Con-sider first the case where the carbon tax or cap-and-trade system is imposed upstream. The carbon tax or allowance price raises costs to suppliers of fossil fuels; the higher costs are likely to be reflected to some extent in the prices charged to users of these fuels. This generates incentives for firms to conserve on the use of these fuels in the production process. Because emis-sions pricing will raise the price of coal (the most carbon-intensive fossil fuel) relative to the prices of oil and natural gas, it will encourage produc-ers that have a choice among fossil fuels to shift their demands away from coal and toward oil or natural gas. Electric utilities, in particular, will be encouraged to shift their demands from coal-generated to oil- or natural-gas-generated electricity in the wholesale market as the increased price of coal will lead to higher prices of electricity produced from coal. These changes in the fuel mix imply reduced CO_2 emissions and help firms mini-mize their tax obligations or the number of allowances they would need to hold. In addition, since the higher fuel prices will be shifted downstream to some degree, emissions pricing will cause the prices of carbon-intensive goods and services—that is, goods and services for which carbon-based fuels represent a significant share of the value of inputs—to rise relative to the prices of less carbon-intensive goods and services. As a result, it will cause demands by consumers to shift away from carbon-intensive goods and services, which helps reduce CO_2 emissions.

All these channels also apply when the points of regulation for the car-bon tax or cap-and-trade program are further downstream. Suppose, for example, emissions pricing is applied midstream—on refiners based on the carbon content of the fuels they purchase and on generators based on the carbon content of their fuels. In this case the tax will again encourage refiners and utilities to engage in fuel switching and consumers of refined products or of electricity to reduce their demands for these higher-priced goods and shift toward other, less carbon-intensive goods and services. When the price is applied downstream at the point of combustion, households have an incentive to substitute or reduce consumption of natural gas and gasoline. In both the midstream and downstream regulations, the reduced demands for fossil fuels will imply reduced supplies of these fuels by the producers upstream. As a result, even though the tax is not imposed directly on the upstream suppliers of fossil fuels, it will have similar impacts on upstream suppliers to those of an upstream tax.[2]

Attractions of Emissions Pricing

Emissions pricing is just one of many options for reducing CO_2 emissions. Alternatives include mandated low-carbon technologies or equipment such as the requirement that coal and natural gas electricity generators install carbon capture and storage (CCS) technology, and performance standards such as the Corporate Average Fuel Economy (CAFE) standards applied to automobile manufacturers. CAFE standards reduce gasoline consumption per mile and thereby imply lower emissions of CO_2 (and other pollutants) per mile as well.

Economists point out several advantages of emissions pricing over these alternatives, advantages related to the ability to achieve emissions reductions at low cost.[3] One advantage is flexibility. Rather than require a particular way to reduce emissions, emissions pricing gives firms flexibility to find the lowest-cost way to achieve the reductions.

A second advantage of emissions pricing is its potential to produce an especially cost-effective allocation of abatement efforts across firms. If broad-based, the emissions price will apply uniformly to all purchasers of fossil fuels or their downstream products. A cost-minimizing competitive firm will reduce emissions up to the point where, at the margin, the costs of abatement equal the avoided tax. Hence uniform emissions pricing promotes equality of marginal abatement costs across firms—a condition for minimizing the aggregate costs of emissions abatement. As the environmental economics literature has emphasized,[4] under conventional regulation it is difficult, if not impossible, to bring about equality of marginal abatement costs across firms: regulators simply do not have sufficient information to set the regulations at levels that would bring this about. The marginal abatement costs depend on the specific features of production processes, features that differ across all the facilities covered by the regulation. Regulators do not have information at this level of detail.

A third advantage is that emissions pricing tends to encourage more efficient use of demand-side conservation than that of conventional regulations. Under conventional regulations, firms are compelled to reduce emissions but are not charged for whatever emissions they continue to generate after they have undertaken some abatement. As indicated in various studies,[5] this implies that prices of output are below the levels that would bring about the most efficient amount of demand-side conservation. To reduce emissions, firms rely principally on input substitution or end-of-pipe treatment; without the sufficient increases in prices of output, consumers of energy goods such as electricity do not face the full incentive to reduce their consumption.[6] In contrast, emissions pricing exploits the demand-conservation channel by yielding output prices closer to the efficiency-maximizing level.[7]

Finally, emissions pricing can promote cost-effectiveness through the recycling of any revenues the policy might generate. This is a potential advantage over conventional regulations, which generally do not bring in revenues.[8] Whether this advantage is realized depends on the form of revenue recycling. As discussed in chapter 2, exploiting this advantage requires that the revenues be recycled through cuts in the rates of preexisting taxes, rather than through fixed rebates, since cutting prior tax rates reduces some of the excess burden from these taxes. Costs are reduced most when the recycling is targeted toward cutting the rates of prior taxes that are exceptionally distortionary.

Of course, cost-effectiveness is not the sole consideration in evaluating the relative merits of emissions pricing and other climate policy options. Fairness is another key consideration. The costs from emissions pricing can be distributed unevenly across firms, households, or regions. For example, the coal mining industry and the households and regions that are supported by coal mining can be particularly impacted. And low-income households that spend a large fraction of income on carbon-intensive goods such as home heating or transportation can suffer a disproportionate cost relative to income. Groups facing significant costs can and do exert considerable opposition. Whether emissions pricing is imposed through a carbon tax or via cap and trade, achieving a design that helps avoid a highly uneven distribution of the costs has attractions in terms of fairness and political feasibility.

In this chapter we focus mainly on the aggregate costs of emissions pricing, starting with those under a carbon tax and then turning to those under cap and trade. Chapter 7 complements this chapter's focus by considering how the aggregate costs are distributed across industries and households under various policy designs. It will show that it is possible to recycle the revenues in a way that avoids uneven impacts on industry profits and/or avoids potential regressivity of the tax. In some cases, smoothing the impacts through revenue recycling raises the cost in the aggregate: there can be a trade-off between efficiency and fairness. We refer briefly to some trade-offs later in this chapter and consider them more extensively in chapter 7.

THE CARBON TAX

Modeling the Carbon Tax

In simulations of the carbon tax, the tax applies to the quantity of fossil fuel input to refiners, electricity generators, and all other direct purchasers of fossil fuels. The tax is the product of a carbon tax rate, expressed as dollars per ton of a given fossil fuel, and a carbon coefficient, defined as tons of carbon per unit of the fossil fuel. For example, 0.43 tons of CO_2 emissions are associated with a barrel of oil; therefore, a \$10 per ton carbon tax amounts to a tax of \$4.30 on each barrel of oil.

As indicated in chapter 3, the carbon tax raises the input price for industries that use fossil fuel inputs. Specifically, the price of domestically produced fossil fuel input i to industry j can be expressed as: $p_{ij}^d = p_i(1+\tau_{ij}^I) + d_j c_{ij} p^{co_2}$, where p_i is the before-tax producer price of input i, τ_{ij}^I is any existing ad valorem tax on input i paid by industry j, p^{co_2} is the carbon tax per (metric) ton of emissions, c_{ij} is the carbon coefficient that converts inputs of good i into emissions from industry j, and d_j is a dummy variable equal to 1 if industry j is covered. An analogous equation for p^f_{ij} applies to the price of foreign-produced fossil fuels. (See appendix B for details.) Table 4.6 indicated the carbon coefficients employed and the sources of these coefficients.

The increased fuel prices under a carbon tax induce behavioral changes along several dimensions. First, because the carbon coefficients differ across fuels, the tax affects fuel prices differently. This prompts firms to substitute toward the fuels whose relative prices have not risen. Second, because the tax also raises the price of carbon-based energy relative to non-carbon-based energy sources and relative to nonenergy inputs, it promotes substitutions away from the carbon-based energy sources. Third, higher-priced energy influences the costs of production, which affects the relative demands for the outputs of various industries. Finally, higher energy prices also influence the composition and levels of investment, both by affecting the costs of producing capital goods and by influencing demands for future outputs and the potential dividends from current investments. For example, emissions pricing will alter the absolute and relative returns to investments in various forms of energy generation. It will tend to promote increased investment in low-carbon generation while leading to lower investment in carbon-based (fossil) generation. These changes in investment will affect future capacity and the extent to which utilities can substitute away from dirty generation in the future.

The preceding discussion focuses on the higher prices from a carbon tax and the costs that such a tax might impose on the economy. This could give the impression that a carbon tax mainly involves losses. But it's important to recognize (as indicated in chapter 2) that under some circumstances a carbon tax can increase income (i.e., involve negative costs) *even before considering the economic benefits (avoided damage) implied by an improved environment*. Moreover, once one accounts for the environment-related economic benefits, the prospects for overall benefits to the economy and well-being are very good. According to the theory presented in chapter 2, an appropriately scaled carbon tax offers net benefits once the environment-related economic benefits are taken into account. Numerical simulation results described later in this book bear this out.

Some designs of carbon tax policies involve recycling of revenues to the private sector, either to firms or to households. Recycling to firms can take the form of cuts in the corporate tax rate or lump-sum tax credits. The former correspond to a reduction in τ^a, while the latter are represented as an increase in *LS*. (Appendix B provides details.) Each of these forms of recycling contributes positively to profits. Recycling to the household can be accomplished through lump-sum tax rebates or cuts in individual income tax rates. With lump-sum rebates, households receive positive rebates G_t^c.[9] Cuts in direct labor taxes τ^L increase after-tax labor income $(1-\tau^L)w_t l_t$ in the household budget constraint (equation [3.11]). Reduced tax rates on dividend income, capital gains, or interest income (τ^e, τ^v, and τ^b) increase $\bar{r}$, the after-tax return to capital.

Design Dimensions

Several features comprise the design of a particular carbon tax policy. One important dimension is the tax's *time profile*—the initial tax rate (defined in dollars per ton of CO_2 equivalent) and the change in that rate over time. The ideal initial rate depends on several considerations. In the absence of other taxes or economic distortions, the theoretically optimal (efficiency-maximizing) environmental tax rate is the marginal external cost (climate-related damage) from emissions. However, as indicated in chapter 2, in a real-world setting where governments rely to some extent on distortionary taxes, the gross costs of an environmental tax tend to be higher, a reflection of the tax-interaction effect. Accordingly, the theoretically optimal environmental tax rate is lower: it is the marginal external cost (climate-related damage) divided by the marginal cost of public funds (MCPF).[10] There is considerable uncertainty surrounding the magnitude of the marginal external cost, which is often termed the *social cost of carbon* (SCC). Recent estimates suggest that the SCC would increase over time with rising concentrations of greenhouse gases.[11] This implies that the optimal carbon tax time profile is one that rises over time. Further considerations might warrant a gradual initial ramp-up of the carbon tax rate, so that it reaches the ratio of marginal external cost to the MCPF only after a few years. The gradual phase-in facilitates smoother capital and labor adjustments and thereby helps avoid the significant macroeconomic impacts that faster implementation could produce.

A second dimension is *breadth of coverage*. As indicated earlier, cost-effectiveness is generally enhanced when the coverage of the tax is broader. In our carbon tax simulations, we focus mainly on an economywide policy, one that covers all sources of carbon dioxide emissions from burning fossil fuels. However, we also consider a policy that covers emissions from the electric power sector only.

A third dimension is the *point of regulation*: upstream, midstream, or downstream. Even if, under some circumstances, this location does not affect the ultimate economic outcomes (see footnote 2 earlier), the choice can be important politically, as politicians and the general public may expect very different impacts depending on where the tax is imposed along the supply chain. Consumers sometimes object more to the taxes they pay directly than to taxes on producers that, because of producers' ability to pass through the costs, can affect consumers just as much. This may reflect the fact that consumers do not fully recognize the potential for pass-through or the related fact that the costs in this latter case are less visible.

A fourth dimension is the extent and nature of *revenue recycling*. Revenues from a carbon tax can be used to finance increases in government spending or can be recycled to the private sector. Under a revenue-neutral carbon tax policy, revenues from the policy are recycled to the private sector so that the total taxes paid by the private sector are not changed and hence the government budget is not expanded. It is important to note that the revenue yield of the carbon tax is not simply the carbon taxes paid. As discussed in chapter 2, a carbon tax gives rise to a tax-base effect, altering the base of other taxes and the revenues they generate. It generally affects, in particular, the tax bases of payroll, individual income, and corporate income taxes, and thus the revenues generated by those taxes. Thus, the revenue impact of the carbon tax (before recycling) is the carbon taxes paid minus any policy-induced loss of revenue from existing taxes. Revenue neutrality requires that the recycled revenues equal this *net* revenue. As indicated earlier, this net revenue can be returned in lump-sum fashion or used to finance cuts in the rates of preexisting distortionary taxes.

Central Case Specifications

Our simulations consider a range of specifications for each of these design dimensions. The simulations revolve around a "central case." We choose features of this case with an eye to cost-effectiveness. For example, the stringency of the tax in the central case is broadly consistent with the level

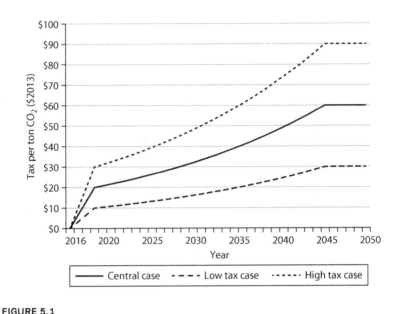

FIGURE 5.1

Alternative Carbon Tax Profiles, 2016–2050

that would be justified by recent estimates of the social cost of carbon. The central case has the following features:

- *Time profile:* Figure 5.1 displays the central case time profile of the carbon tax as well as higher tax and lower tax variants. In all these cases, the tax is imposed in the year 2017, assumed to be unanticipated, and phased in over three years in equal increments. In the central case, the tax reaches $20 per ton (in 2013 dollars) by 2019; hence the rates are $6.67, $13.33, and $20.00 in 2017, 2018, and 2019, respectively. After 2019, the tax rate is increased in real terms at 4 percent annually. We cap the tax when it reaches a real rate of $60 per ton. It reaches that rate in 2048 and is kept at that rate from then on. In alternative cases, we maintain the three-year, equal-increment phase-in of the tax as well as the subsequent 4 percent annual growth rate in the tax rate. What differs is the height of the time profile (starting with the 2019 tax rate) and the maximum tax rate that is reached after several decades. Figure 5.1 presents two examples: a high tax case, with a post phase-in (2019) tax rate of $30 per ton, and a low tax case, with a $10 per ton tax post phase-in. In these alternative cases, the maximum tax rate is set so that the ratio of this rate to the 2019 tax rate is the same as the ratio in the central case. Thus the maximum rate is $90 per ton in the high tax case and $30 per ton in the low tax case.

- *Coverage:* The base of the tax includes all industrial purchasers of fossil fuels. Our central case policy also covers imports of carbon-based refined products such as gasoline, diesel, and heating oil. Tariffs equal to the carbon content of refined products ensure that these emissions are covered. This specification implies coverage of 99.9 percent of emissions from the burning of fossil fuels and 94 percent of all U.S. carbon dioxide emissions.[12]

- *Points of regulation:* Our central case considers a carbon tax imposed midstream at the industrial user's gate. Because the tax is based on carbon content, it reflects the CO_2 released not only by the industrial purchaser that combusts the fuel but also by the combustion of any downstream products that the purchaser might produce. This method is relatively straightforward to administer, as most purchasers of fossil fuels are already taxed entities. It can also involve lower administrative costs than a more upstream approach, since there are fewer (midstream) purchasers of fuels than (upstream) suppliers; for example, there are far fewer refineries than oil wells in the United States. In our model, this method allows flexibility in altering the coverage of the policy.[13]

- *Revenue recycling:* We focus on carbon tax policies that are revenue-neutral as defined earlier: whatever revenues might be generated by the policy in question (net of any loss of revenue associated with the tax-base effect) are returned to the private sector in the form of reduced tax rates or lump-sum rebates. Recycling takes one of four forms in our central case simulations: (1) lump-sum rebates, (2) cuts in the employee payroll tax rate, (3) cuts in individual income tax rates,[14] and (4) cuts in corporate income tax rates.

Central Case Results

Here we focus on the impacts of our central-case specification for the carbon tax. We consider both economywide outcomes, such as the effect on nationwide emissions and GDP, as well as the impacts across industries and on the representative household. Later we consider a wide range of carbon tax stringencies. Chapter 7 will extend the analysis by examining how the different industry impacts can be modified through various compensation schemes. That chapter will also assess the impacts on different household income groups and consider a range of compensation approaches that alter the distribution of household impacts.

Emissions, Revenues, and Tax Rates

Figure 5.2 presents the time profile for CO_2 emissions in the reference case and in the carbon tax's central case, when tax revenues are recycled in lump-sum fashion. The emissions time profile is similar in the other recycling cases.[15] By 2020 and 2035, emissions are reduced by approximately 17 and 30 percent, respectively, from the reference case value. Emissions continue to decline both absolutely and relative to the reference case up to the year 2050, at which point they are about 37 percent below the reference case level for that year. From the year 2050 on, emissions grow

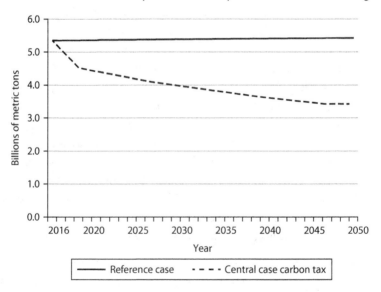

FIGURE 5.2

Economywide Carbon Dioxide Emissions, 2016–2050

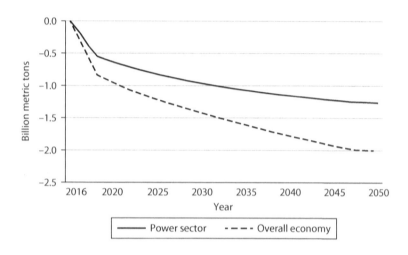

FIGURE 5.3

Change in Carbon Dioxide Emissions, Central Case Carbon Tax, 2016–2050

at roughly the rate of growth of the overall economy, remaining about 38 percent below the reference case levels for corresponding years.

A large fraction of the economy's emissions reductions come from the electric power sector. This is shown in figure 5.3, which displays the power sector's reductions as well as the overall economy's reductions. Over the interval 2017–2050, this sector accounts for 64–68 percent of emissions reductions in each period. This sector's reductions are accomplished through substantial switching away from coal-fired generation toward natural-gas-fired and renewables-based generation.

Figure 5.4 displays the time profile of the carbon tax's gross and net revenues, when the revenues are recycled through lump-sum rebates to the household. Notably, net revenues (which account for the tax-base effect described earlier) are significantly below gross revenues. In fact, they are virtually zero (actually, slightly negative) in 2017, the first year of the policy. This reflects the assumption that the carbon tax is unanticipated. Because it is not anticipated, it yields significant windfall losses in 2017— the year the policy is "revealed"—in the form of reductions in the value of carbon-related firms.[16] Stockholders can deduct such capital losses from their taxable income. The associated reduction in overall capital gains tax revenues in 2017 fully offsets the carbon tax's gross revenues of $37 billion.

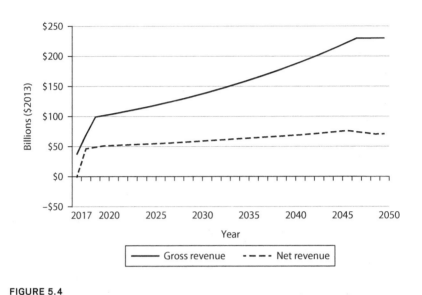

FIGURE 5.4

Carbon Tax Revenue, Central Case, 2016–2050

This significant offset to revenues is confined to the year 2017, when the windfall losses from the carbon tax "surprise" occur. After 2017, net revenues are positive. In 2019 (when the price hits $20 per ton), gross and net revenues are approximately $99 billion and $49 billion, respectively. After that year, gross revenues continue to grow while net revenues remain approximately constant.

Table 5.1 indicates the extent to which marginal rates on payroll, individual, and corporate income taxes change when carbon tax revenues finance cuts in those rates. The table includes the unchanged rates when recycling is carried out through lump-sum rebates, along with the changed rates in the other recycling cases. The figures in parentheses are the percentage point reductions in these rates from their reference case values. The revenues from the central case carbon tax are able to finance considerably larger cuts in the corporate income tax rate than in individual income tax rates, in keeping with the fact that the base of the corporate income tax is much smaller.

TABLE 5.1 Marginal Tax Rates After Recycling of Carbon Tax Revenues

	RECYCLING METHOD			
TAX	LUMP-SUM REBATES	CUTS IN EMPLOYEE PAYROLL TAXES	CUTS IN INDIVIDUAL INCOME TAXES	CUTS IN CORPORATE INCOME TAXES
Individual—labor	26.78	26.78	26.38	26.78
	(0.00)	(0.00)	(−0.39)	(0.00)
Individual—dividends	27.73	27.73	27.41	27.73
	(0.00)	(0.00)	(−0.32)	(0.00)
Individual—interest	31.78	31.78	31.41	31.78
	(0.00)	(0.00)	(−0.37)	(0.00)
Individual—capital gains	26.12	26.12	25.81	26.12
	(0.00)	(0.00)	(−0.31)	(0.00)
Payroll (employee share)	6.95	6.51	6.95	6.95
	(0.00)	(−0.44)	(0.00)	(0.00)
Corporate	40.00	40.00	40.00	35.23
	(0.00)	(0.00)	(0.00)	(−4.77)

Note: Numbers in parentheses represent percentage point changes in tax rates from reference case values.

Despite a relatively large difference between gross and net revenues, the policy provides enough net revenue to finance notable reductions in preexisting taxes on capital or labor income. Which preexisting taxes are cut significantly affects the policy costs, as discussed later.

Impacts on GDP, GDP Components, and Welfare

Here we consider several economywide impacts, disregarding for now the benefits associated with avoided climate-related damage. (We address the benefits later.) As described in chapter 2, emissions pricing imposes costs on an economy in the form of the primary cost (or direct cost) of abatement and the tax-interaction effect, but offsets some or all of these costs via the revenue-recycling effect. In the E3 model we measure costs through changes in GDP and consumption, and through welfare impacts as measured by the equivalent variation.[17] Figures 5.5*a*, *b*, and *c* show the impacts of the carbon tax on real GDP and its consumption and investment components, for cases involving lump-sum rebates, payroll tax recycling, individual tax recycling, and corporate tax recycling.[18] We do not display the impacts on real government spending, since all policy experiments maintain such spending at reference case levels.

The GDP impact depends significantly on the type of revenue recycling. As shown in figure 5.5a, using the revenues to finance cuts in marginal tax rates significantly reduces this cost relative to the case in which revenues are recycled through lump-sum rebates. The distortionary cost or excess burden of taxes is a function of marginal tax rates. When carbon tax revenues are devoted to marginal rate cuts, some of this distortionary cost is avoided. In the year 2030, for example, the GDP costs under payroll, individual, and corporate tax rate recycling are 25, 32, and 79 percent smaller, respectively, than under lump-sum recycling. The

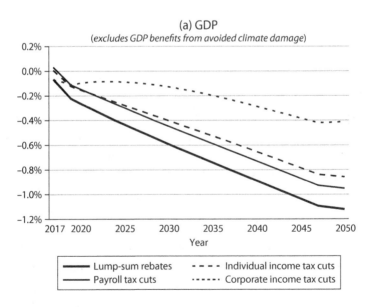

FIGURE 5.5

Percent Change in GDP, Consumption, and Investment, 2017–2050

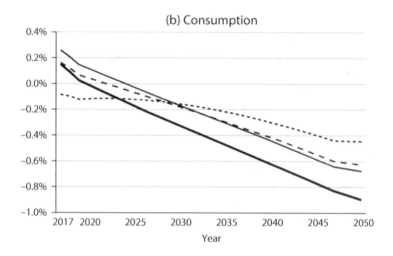

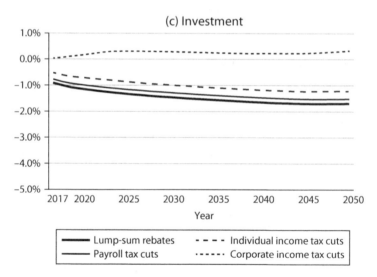

FIGURE 5.5 *(Continued)*

rankings of cost savings match the rankings of the marginal excess burdens of the different taxes involved: as indicated in chapter 4, the marginal excess burdens of the payroll, individual, and corporate taxes are 0.24, 0.42, and 1.08, respectively, in the model.

The shapes of the time profiles of GDP impacts are similar under lump-sum, payroll tax, and individual income tax recycling. For those cases, the percentage reductions in GDP tend to increase over time. Under corporate income tax recycling, however, the pattern of impacts is different in two ways. First, the magnitude of the impacts is smaller. Because the corporate tax is more distortionary (has a larger marginal excess burden) than both the payroll tax and the individual tax, it produces larger cost savings when carbon tax revenues are devoted to cutting the marginal rates of this tax. In addition, the percentage change in GDP changes by less over time. This reflects the positive impact of the corporate tax cut on investment, which helps minimize adverse effects on the economy's capital stock and thereby helps prevent losses in GDP. The curvature in the GDP time profile under corporate rate cuts is primarily due to the impacts of the emerging backstop industry, which begins producing in 2023.

Readers may wish to know how the E3 model's results for GDP compare with those from other models that consider an economywide carbon tax. Interpreting differences in results is often difficult because the models often focus on policies that differ in terms of stringency, breadth, or other design dimensions, making apples-to-apples comparisons impossible. Fortunately, a recent study by Stanford University's Energy Modeling Forum (EMF) provides information that facilitates a comparison of results. The study, "EMF 32, An Inter-model Comparison of US Greenhouse Gas Reduction Policy Options," was conducted in 2016 and 2017. As part of this study, E3 and several other models assessed the impacts of an economywide carbon tax starting at $25 (in 2010 dollars) in 2020 and increasing at 5 percent per year until the year 2050. The models considered this carbon tax time profile under three common recycling scenarios. Although the common carbon tax time profile in the EMF study does not perfectly match that of our central case carbon tax, the models' results from this common policy scenario allow for useful comparisons.

Table 5.2 displays the GDP results from six models that focused on this scenario. In addition to E3, the models are DIEM from Duke University's

TABLE 5.2 Cross-Model Comparison of Percentage Changes in GDP (Relative to Baseline)

MODEL	2020			2030			2040		
	LUMP-SUM REBATES	LABOR INCOME TAX CUTS	CAPITAL INCOME TAX CUTS	LUMP-SUM REBATES	LABOR INCOME TAX CUTS	CAPITAL INCOME TAX CUTS	LUMP-SUM REBATES	LABOR INCOME TAX CUTS	CAPITAL INCOME TAX CUTS
E3	-0.2	-0.2	-0.3	-0.8	-0.7	-0.6	-1.4	-1.3	-1.1
DIEM	-0.4	-0.2	0.4	-0.4	-0.2	0.8	-0.6	-0.4	0.9
IGEM	-0.5	-0.1	0.3	-0.8	0.2	0.5	-1.1	0.2	0.5
NewERA	-0.3	-0.2	0.1	-0.5	-0.4	0.2	-0.7	-0.6	0.2
RTI-ADAGE	-0.6	-0.4	0.6	-0.8	-0.6	0.9	-1.1	-0.8	1.0
ReEDS-USREP	-0.5	-0.3	-0.4	-0.3	-0.1	0.0	-0.2	0.1	0.2
Model average	**-0.4**	**-0.2**	**0.1**	**-0.6**	**-0.3**	**0.3**	**-0.8**	**-0.5**	**0.3**

Nicholas Institute for Environmental Policy Solutions (Ross, Fawcett, and Clapp 2009); IGEM, featured in Jorgenson et al. (2013); the NewERA model from NERA Economic Consulting (featured in Smith et al. 2013); RTI-ADAGE from RTI International (Ross 2014); and ReEDS-USREP, a joint collaboration between the National Renewable Energy Laboratory (NREL) and the MIT Joint Program on the Science and Policy of Global Change.

The different models all tend to find that the GDP impacts are most favorable (or least costly) when recycling is in the form of capital income tax cuts and least favorable under lump-sum recycling. The impacts under labor income tax cuts are in the middle.

In 2020, the year of policy implementation in the EMF scenario, the E3 model yields smaller, average, or larger adverse GDP impacts than the other models under lump-sum rebates, labor income tax cuts, and capital income tax cuts, respectively. E3 leads to a more pessimistic outcome in the case of capital income tax cuts. This may reflect the presence of capital adjustment costs in E3, a unique feature of our model. These adjustment costs slow the ability of the capital tax cuts to improve efficiency through the reallocation of capital.

Over the longer term, the E3 model's GDP impacts tend to be less optimistic than those from the other models. Also, there is less variation in GDP impacts across the alternative recycling methods in E3 than in the other models. The other models suggest that under capital income tax recycling, the long-run impact on GDP is positive. E3 does not generate the positive GDP impact under the EMF's carbon tax time profile or in the time profile of our central case, although we do obtain it when the carbon tax rate is lower, as discussed later. One contributor to E3's less optimistic results is that the preexisting tax system is more distortionary in E3 than in other models, a function of E3's greater tax detail. E3 distinguishes several types of taxes on labor (wage income taxes and payroll taxes) and capital (corporate, dividends, capital gains, and interest); most other models only include wage income taxes and average capital taxes. Because the distortionary cost of taxes is a convex (superlinear) function of tax rates, the averaging of other taxes in other models implies that the preexisting tax system is less distortionary; this in turn implies a lower tax-interaction effect, which implies lower economic costs.

Figures 5.5b and c above reveal the E3 central case carbon tax impacts on consumption and investment. The time profiles for consumption have a similar shape for the cases other than corporate income tax recycling. In these cases, the policy initially yields higher consumption but ultimately implies lower consumption than in the reference case. In contrast, in the case of corporate income tax recycling, there is a larger negative impact on consumption in the short term, but a smaller negative impact in the long run. The negative initial impact on consumption is consistent with the fact that the corporate tax stimulates savings and investment and thereby reduces the share of income devoted to consumption. The longer-term impact on consumption is smaller, in keeping with the higher capital stock and output that stem from cuts in the corporate tax rate.

Table 5.3 offers further information on the policy costs. The top panel expresses the GDP costs as a percentage of reference case GDP and per ton of CO_2 reduced, over the interval 2017–2050. The bottom panel shows the aggregate welfare cost as measured by the equivalent variation.

TABLE 5.3 GDP and Welfare Costs of a Carbon Tax Under Alternative Recycling Methods

| | | RECYCLING METHOD | | |
	LUMP-SUM REBATES	CUTS IN EMPLOYEE PAYROLL TAXES	CUTS IN INDIVIDUAL INCOME TAXES	CUTS IN CORPORATE INCOME TAXES
GDP Costs[a]				
—As percent of reference GDP	0.64	0.49	0.45	0.19
—Per ton of CO_2 reduced[b]	$98.02	$76.18	$69.32	$30.83
Welfare costs[c]	$3,356.47	$2,844.71	$2,479.64	$1,068.60
—As percent of wealth	0.41	0.35	0.30	0.13
—Per dollar of gross revenue	$0.43	$0.36	$0.31	$0.13
—Per ton of CO_2 reduced	$42.12	$35.80	$31.30	$13.95

[a]GDP costs measured as present value of real GDP loss, 2017–2050, using 3 percent real interest rate.
[b]Present value of cumulative tons reduced, using 3 percent real interest rate.
[c]Welfare costs are the negative of the equivalent variation, expressed in billions of 2013 dollars.

The pattern for the welfare costs resembles the pattern for GDP costs. As with the GDP costs, the welfare costs are considerably smaller when revenues are recycled through cuts in marginal rates and the revenue-recycling effect is exploited, compared with the case of recycling via lump-sum rebates. Relative to the case with lump-sum recycling, the policies with recycling through cuts in the rates of corporate taxes, individual income taxes, and payroll taxes involve welfare costs that are 67, 26, and 15 percent lower, respectively. Again the ranking matches the ranking of the MEB values of these taxes.

Two key takeaways emerge from our analysis of the gross costs of reducing emissions through a carbon tax. First, the choice of how to use the revenues matters significantly, both to the general level of cost and to the shape of the cost-time profile. Second, as can be seen from figure 5.5a and table 5.3, in the central case the model does not yield a double dividend under any of the forms of recycling considered. The revenue-recycling effect is not large enough to offset the combination of primary cost and the tax-interaction effect. Later, we revisit the prospects for a double dividend by considering a range of carbon tax time profiles and alternative parameter assumptions.

Impacts on Industry Profits, Prices, Output, and Labor Input

One would expect emissions pricing to have the largest impacts on prices, output, and profits in the industries with the highest carbon intensities. Our model enables us to go beyond these qualitative predictions by providing quantitative assessments of the magnitudes involved.

We start with the profit impacts, which are displayed in table 5.4. The coal mining industry and the coal-fired electricity generators—industries with exceptionally high carbon intensities—experience the most significant losses of profit. Although the coal mining industry does not directly pay the carbon tax, it bears a significant burden from the tax as a result of the policy-induced reduction in demand for coal.[19] Significant profit losses also occur to other-fossil generators and in the electricity transmission and distribution and natural gas distribution industries, which rely significantly on carbon-intensive fuels. Railroad transportation also experiences profit losses because of the reduced demand stemming from

TABLE 5.4 Impacts on Industry Profits

Percentage Changes in the Present Value of Profits over the Infinite Horizon

INDUSTRY	LUMP-SUM REBATES	CUTS IN EMPLOYEE PAYROLL TAXES	CUTS IN INDIVIDUAL INCOME TAXES	CUTS IN CORPORATE INCOME TAXES
		RECYCLING METHOD		
Oil extraction	−0.1	−0.1	−0.1	6.8
Natural gas extraction	−23.5	−23.4	−23.3	−20.3
Coal mining	−45.9	−45.8	−45.7	−45.7
Electric transmission and distribution	−7.9	−7.7	−7.6	−5.5
Coal-fired electricity generation	−74.7	−74.6	−74.6	−75.0
Other-fossil electricity generation	−18.5	−18.3	−18.3	−14.8
Nonfossil electricity generation	62.7	63.0	63.4	66.1
Natural gas distribution	−8.4	−8.2	−8.1	−5.7
Petroleum refining	−6.3	−6.2	−6.1	−3.2
Pipeline transportation	−7.2	−7.1	−7.0	−3.3
Mining support activities	−5.5	−5.3	−4.9	−0.5
Other mining	−3.2	−3.0	−2.8	−0.2
Farms, forestry, fishing	−1.8	−1.6	−1.6	1.3
Water utilities	−1.0	−0.8	−0.8	1.2
Construction	−2.3	−2.1	−1.8	0.5
Wood products	−2.0	−1.9	−1.7	0.7
Nonmetallic mineral products	−2.3	−2.2	−2.0	0.5
Primary metals	−3.3	−3.2	−3.1	−0.8
Fabricated metal products	−2.1	−1.9	−1.8	0.2
Machinery and misc. manufacturing	−1.9	−1.7	−1.6	0.8
Motor vehicles	−1.6	−1.4	−1.3	0.7
Food and beverage	−1.6	−1.4	−1.4	1.3
Textile, apparel, leather	−1.7	−1.4	−1.4	1.7
Paper and printing	−1.8	−1.6	−1.6	0.2
Chemicals, plastics, and rubber	−2.7	−2.5	−2.4	−0.2
Trade	−1.6	−1.4	−1.4	1.1
Air transportation	−2.8	−2.6	−2.6	−0.4
Railroad transportation	−3.6	−3.5	−3.4	0.3
Water transportation	−2.4	−2.3	−2.2	−0.2
Truck transportation	−2.0	−1.8	−1.8	0.1
Transit and ground passenger transportation	−1.2	−1.0	−1.0	1.4
Other transportation and warehousing	−1.8	−1.7	−1.7	0.8
Communication and information	−1.1	−0.9	−0.9	1.7
Services	−1.2	−1.0	−1.0	0.4
Real estate and owner-occupied housing	−1.1	−0.9	−0.9	−0.4
All industries[a]	−1.3	−1.1	−1.1	0.2

[a]Weighted average, based on 2013 output levels.

a decrease in the amount of coal that needs to be transported from the mine to the generator. Carbon-intensive industries face large profit losses regardless of the recycling method, but some low-carbon industries such as trade, food and beverage manufacturing, and communications have positive profit impacts when recycling takes the form of a cut in the corporate income tax rate.

Impacts on producers' nominal output prices are displayed in table 5.5. The table shows percentage changes from the reference case for the years 2020, 2035, and 2050 for the case involving recycling via lump-sum rebates.[20] The price changes reflect the higher prices of fossil fuel inputs. The price impacts are largest in the industries that have the greatest intensity of use of fossil fuels: coal-fired generation, other fossil (principally natural gas) generation, electricity transmission and distribution, and natural gas distribution. The reduction in the producer price of coal reflects the general equilibrium nature of the model. Purchasers of coal see their price rise because it includes the carbon tax; hence they demand less coal. This drop in demand results in a decrease in the producer price received by coal mining firms.[21]

The impacts on industry output, shown in table 5.6, parallel the impacts on profits shown in table 5.4. Coal mining and coal-fired electricity experience sharp drops in output. The major suppliers and users of natural gas—the natural gas extraction industry, other-fossil electricity generators, and the natural gas distribution industry—also see declines in output, though the impacts are smaller than those experienced by the coal industry, reflecting the lower carbon coefficient of natural gas relative to coal.

As shown in table 5.6, the carbon tax has relatively little impact on the output of the domestic oil extraction industry. Chapter 3 indicated that the E3 model regards imported oil as the marginal source of supply of oil to meet domestic demand. As a result, the reduction in the demand for oil is accommodated mainly through reduced demand for imports rather than reduced domestic production of oil.

Differences in recycling methods do not much change the variation in impacts across industries; such variation mainly stems from differences in carbon intensities. For the industries that experience adverse impacts,

TABLE 5.5 Impacts on Producer Prices

Percentage Changes from Reference Case Values

INDUSTRY	2020	2035	2050
Oil extraction	0.6	1.8	2.6
Natural gas extraction	−5.9	0.9	0.7
Coal mining	−9.5	−1.5	0.7
Electric transmission and distribution	6.8	12.7	15.7
Coal-fired electricity generation	25.5	61.7	97.7
Other-fossil electricity generation	12.9	24.6	36.5
Nonfossil electricity generation	9.6	5.1	3.6
Natural gas distribution	4.2	12.2	18.8
Petroleum refining	7.3	12.4	17.6
Pipeline transportation	0.8	3.6	5.1
Mining support activities	−0.3	0.7	1.0
Other mining	0.5	1.8	2.4
Farms, forestry, fishing	0.4	1.4	1.9
Water utilities	0.3	0.7	1.0
Construction	0.4	0.9	1.2
Wood products	0.3	1.0	1.3
Nonmetallic mineral products	0.3	1.2	1.7
Primary metals	0.9	2.2	2.7
Fabricated metal products	0.3	0.9	1.2
Machinery and misc. manufacturing	0.0	0.6	0.8
Motor vehicles	0.1	0.7	0.9
Food and beverage	0.3	1.1	1.4
Textile, apparel, leather	0.2	0.7	1.0
Paper and printing	0.4	1.1	1.5
Chemicals, plastics, and rubber	0.9	2.3	3.0
Trade	0.1	0.5	0.6
Air transportation	1.3	2.7	3.7
Railroad transportation	−0.7	0.7	1.4
Water transportation	1.2	2.4	3.2
Truck transportation	1.2	2.3	3.0
Transit and ground passenger transportation	0.7	1.3	1.8
Other transportation and warehousing	0.4	1.0	1.3
Communication and information	0.0	0.3	0.5
Services	0.1	0.4	0.5
Real estate and owner-occupied housing	0.3	0.7	1.1
All industries (producer price index)	0.6	1.6	2.3

TABLE 5.6 Impacts on Industry Output

Percentage Changes from Reference Case Values

INDUSTRY	2020		2035	
	LUMP-SUM REBATES	CUTS IN INDIVIDUAL INCOME TAXES	LUMP-SUM REBATES	CUTS IN INDIVIDUAL INCOME TAXES
Oil extraction	−0.5	−0.5	−0.4	−0.3
Natural gas extraction	−6.1	−6.0	−17.3	−17.1
Coal mining	−20.2	−20.1	−39.4	−39.3
Electric transmission and distribution	−5.0	−4.9	−9.0	−8.7
Coal-fired electricity generation	−28.9	−28.8	−61.0	−60.9
Other-fossil electricity generation	−2.2	−2.2	−14.9	−14.6
Nonfossil electricity generation	6.8	6.8	42.0	42.6
Natural gas distribution	−3.2	−3.1	−8.8	−8.6
Petroleum refining	−6.3	−6.2	−10.2	−10.0
Pipeline transportation	−2.4	−2.3	−5.9	−5.7
Mining support activities	−3.2	−2.6	−3.4	−2.8
Other mining	−1.7	−1.5	−3.2	−2.8
Farms, forestry, fishing	−0.9	−0.7	−1.8	−1.5
Water utilities	0.1	0.2	−0.1	0.1
Construction	−0.9	−0.5	−1.4	−0.9
Wood products	−0.9	−0.6	−1.6	−1.2
Nonmetallic mineral products	−1.1	−0.8	−1.9	−1.5
Primary metals	−2.1	−1.9	−3.6	−3.3
Fabricated metal products	−1.0	−0.8	−1.8	−1.5
Machinery and misc. manufacturing	−1.0	−0.8	−1.8	−1.5
Motor vehicles	−0.8	−0.6	−1.3	−1.1
Food and beverage	−0.4	−0.3	−1.2	−0.9
Textile, apparel, leather	−0.6	−0.3	−1.4	−1.0
Paper and printing	−0.8	−0.7	−1.6	−1.4
Chemicals, plastics, and rubber	−1.6	−1.4	−3.0	−2.8
Trade	−0.5	−0.4	−1.0	−0.7
Air transportation	−1.8	−1.7	−3.1	−2.9
Railroad transportation	−1.5	−1.4	−3.2	−3.0
Water transportation	−1.6	−1.5	−2.7	−2.4
Truck transportation	−1.5	−1.3	−2.4	−2.1
Transit and ground passenger transportation	−0.4	−0.3	−0.8	−0.6
Other transportation and warehousing	−1.0	−0.8	−1.6	−1.4
Communication and information	−0.1	0.0	−0.4	−0.2
Services	−0.2	−0.1	−0.4	−0.2
Real estate and owner-occupied housing	−0.1	0.0	−0.4	−0.1
All industries[a]	−0.3	−0.1	−0.6	−0.4

[a] Weighted average, based on 2013 output levels.

the magnitude of these impacts tends to be smaller under marginal tax rate recycling than under lump-sum rebating.

Table 5.7 displays the carbon tax's impacts in the labor market, where labor is measured in effective hours worked.[22] The carbon tax precipitates changes in labor demand that parallel the changes in industry output. The reductions in industry output from a carbon tax exert a negative *scale effect* that works toward a reduction in labor demand. This negative effect overcomes the positive *intensity effect* that results when higher energy prices promote substitutions of labor for energy inputs. In the nonfossil electricity generation industry, the scale effect is positive, reflecting the carbon tax's positive impact on output in this industry. This positive scale effect combines with a positive intensity effect to give rise to increases in labor demand. The services sector experiences a small decline in output from the carbon tax. Under lump-sum recycling, the small negative scale effect is sufficient to produce a small decline in labor demand. In contrast, under individual income tax recycling, the impact on output and the associated scale effect are even smaller, and a positive intensity effect is large enough to overcome the scale effect and yield an increase in labor demand.

We have shown that the carbon tax's impacts on profits, labor demand, and output differ substantially across industries. These differences have important implications in terms of policy fairness and political feasibility. In chapter 7, we look at various ways that some of the tax revenues can be used to minimize industry impacts, and we assess the effects of these alternative revenue approaches on policy costs.

Impacts on Trade Flows

Politicians frequently worry about the impact of a carbon tax on international trade. Opponents of the tax often argue that it will cause production to shift away from the United States to locations without stringent carbon policies, leading to higher imports and lower exports. As Chapter 3 indicated, E3 accounts for potential impacts of domestic policy on the pattern of international trade by incorporating domestic demand for foreign goods (imports) and foreign demand for domestic goods (exports).

The impacts on trade depend on the carbon tax's specific design features. The tax considered here is imposed midstream, that is, on the

TABLE 5.7 Impacts on Labor Demand

Percentage Changes from Reference Case Values

	2020		2035	
INDUSTRY	LUMP-SUM REBATES	CUTS IN INDIVIDUAL INCOME TAXES	LUMP-SUM REBATES	CUTS IN INDIVIDUAL INCOME TAXES
Oil extraction	0.3	0.3	1.5	1.5
Natural gas extraction	−8.5	−8.3	−16.1	−16.0
Coal mining	−23.3	−23.2	−39.6	−39.5
Electric transmission and distribution	1.2	1.4	0.9	1.1
Coal-fired electricity generation	−25.8	−25.7	−57.8	−57.7
Other-fossil electricity generation	−0.2	−0.1	−11.8	−11.5
Nonfossil electricity generation	12.8	13.0	45.5	46.0
Natural gas distribution	1.2	1.3	1.4	1.6
Petroleum refining	−1.5	−1.4	−2.5	−2.3
Pipeline transportation	−0.8	−0.7	−2.3	−2.0
Mining support activities	−3.3	−2.6	−2.9	−2.3
Other mining	−1.3	−1.0	−1.8	−1.5
Farms, forestry, fishing	−0.5	−0.3	−0.8	−0.6
Water utilities	0.3	0.5	0.4	0.6
Construction	−0.7	−0.1	−0.7	−0.2
Wood products	−0.7	−0.3	−0.9	−0.5
Nonmetallic mineral products	−0.8	−0.4	−1.0	−0.7
Primary metals	−1.4	−1.2	−2.1	−1.9
Fabricated metal products	−0.8	−0.6	−1.1	−0.8
Machinery and misc. manufacturing	−1.0	−0.7	−1.4	−1.1
Motor vehicles	−0.7	−0.4	−0.8	−0.6
Food and beverage	−0.2	0.0	−0.5	−0.2
Textile, apparel, leather	−0.4	−0.1	−0.9	−0.6
Paper and printing	−0.5	−0.4	−0.8	−0.6
Chemicals, plastics, and rubber	−0.9	−0.7	−1.4	−1.2
Trade	−0.4	−0.2	−0.6	−0.4
Air transportation	−0.7	−0.6	−1.1	−1.0
Railroad transportation	−1.7	−1.5	−2.6	−2.4
Water transportation	−0.7	−0.5	−0.9	−0.8
Truck transportation	−0.6	−0.4	−0.8	−0.6
Transit and ground passenger transportation	0.1	0.3	0.1	0.3
Other transportation and warehousing	−0.7	−0.5	−0.9	−0.7
Communication and information	−0.1	0.0	−0.2	0.0
Services	−0.1	0.1	−0.1	0.0
Real estate and owner-occupied housing	0.3	0.4	0.3	0.5
All industries[a]	−0.1	0.0	−0.2	0.0

[a] Weighted average, based on 2013 levels of labor demand.

domestic purchasers of fossil fuels. With this design, the tax does not directly tilt the playing field between domestic suppliers of fossil fuels and their foreign counterparts, since the U.S. purchasers of these fuels are taxed irrespective of their origin. Nor does it have any direct effect that would put exports of domestically produced fossil fuels at a disadvantage on the international market, because it is imposed only on primary fossil fuels that are consumed at home.

However, the carbon tax does affect U.S. firms relative to those abroad, especially those firms that utilize large amounts of fossil fuels. Higher fossil fuel prices raise costs to domestic firms that use these fuels directly or indirectly, which tend to raise the prices of domestic goods relative to the prices of the foreign counterparts. In contrast, producers of fossil fuels, with the exception of oil, see their prices fall relative to prices of their foreign counterparts.

The trade patterns shown in tables 5.8 and 5.9 reflect these differing price impacts. As indicated by the far-right column of table 5.8, nearly every industry exhibits an increase in import intensity—the fraction of domestic demand satisfied through imports—as consumers shift away from higher-priced domestic goods and toward lower-priced foreign goods. The changes in import intensity are generally quite small.[23] However, in a significant fraction of industries the *quantity* of imports does not rise, reflecting the countervailing influence of the carbon tax's scale effect—the general reduction in demand that the tax can induce as a consequence of its impact on incomes. In the primary and fabricated metal products industries, for example, the scale effect dominates the intensity (substitution) effect, so that in these industries the quantity of imports declines. In other industries, the intensity effect dominates; for example, imports rise in the food and beverage manufacturing industry.

As shown in table 5.9, exports generally fall in every industry as the foreign demand for domestically produced goods declines, a reflection of the increased relative prices of these goods. The exceptions are in the natural gas extraction and coal mining industries. The carbon tax leads to a *decline* in the relative price of domestically produced natural gas and coal (table 5.5), which induces foreigners to shift their purchases toward the domestically produced fuels.[24]

TABLE 5.8 Changes in Imports in 2020
Under Carbon Tax with Lump-Sum Recycling

INDUSTRY	REFERENCE CASE IMPORTS[a]	PCT CHANGE	REFERENCE CASE IMPORT INTENSITY[b]	CHANGE IN IMPORT INTENSITY
Oil extraction	359.3	−10.9	68.06	−2.45
Natural gas extraction	11.6	−10.7	10.74	−0.24
Coal mining	0.0	−40.3	0.00	0.00
Electric transmission and distribution	2.3	1.6	0.60	0.04
Coal-fired electricity generation	0.0		0.00	0.00
Other-fossil electricity generation	0.0		0.00	0.00
Nonfossil electricity generation	0.0		0.00	0.00
Natural gas distribution	0.0		0.00	0.00
Petroleum refining	82.0	−2.9	12.64	0.24
Pipeline transportation	0.0		0.00	0.00
Mining support activities	1.2	−3.7	0.64	0.00
Other mining	4.4	−0.9	8.53	0.06
Farms, forestry, fishing	51.3	0.1	11.06	0.06
Water utilities	0.0		0.00	0.00
Construction	0.0		0.00	0.00
Wood products	16.4	−0.5	14.67	0.04
Nonmetallic mineral products	21.5	−0.3	17.19	0.08
Primary metals	85.5	−0.8	23.27	0.17
Fabricated metal products	57.6	−0.6	14.78	0.04
Machinery and misc. manufacturing	646.8	−0.5	39.04	0.01
Motor vehicles	237.9	−0.4	31.36	0.02
Food and beverage	70.8	0.8	8.15	0.07
Textile, apparel, leather	139.3	0.4	64.04	0.19
Paper and printing	25.0	−0.3	9.98	0.03
Chemicals, plastics, and rubber	235.9	0.0	21.28	0.19
Trade	0.0		0.00	0.00
Air transportation	27.0	0.3	18.31	0.22
Railroad transportation	0.3	−0.5	0.30	0.00
Water transportation	0.1	0.4	0.16	0.00
Truck transportation	2.3	−0.3	0.84	0.01
Transit and ground passenger transportation	0.0		0.00	0.00
Other transportation and warehousing	0.0	−1.0	0.00	0.00
Communication and information	9.0	0.1	0.76	0.00
Services	159.1	−0.3	1.51	0.00
Real estate and owner-occupied housing	0.0		0.00	0.00

[a]In billions of 2013 dollars.
[b]*Import intensity* is defined as total imports of a good divided by the total domestic consumption of that good.

TABLE 5.9 Changes in Exports in 2020
Under Carbon Tax with Lump-Sum Recycling

INDUSTRY	REFERENCE CASE EXPORTS[a]	PCT CHANGE	REFERENCE CASE EXPORT INTENSITY[b]	CHANGE IN EXPORT INTENSITY
Oil extraction	0.0		0.00	0.00
Natural gas extraction	17.9	6.6	15.64	2.11
Coal mining	13.8	11.8	32.77	13.14
Electric transmission and distribution	1.8	−12.1	0.46	−0.03
Coal-fired electricity generation	0.0		0.00	0.00
Other-fossil electricity generation	0.0		0.00	0.00
Nonfossil electricity generation	0.0		0.00	0.00
Natural gas distribution	0.5	−8.2	0.36	−0.02
Petroleum refining	134.7	−11.8	19.20	−1.13
Pipeline transportation	3.1	−3.5	7.71	−0.09
Mining support activities	3.4	−2.1	1.79	0.02
Other mining	4.4	−3.1	8.53	−0.12
Farms, forestry, fishing	70.3	−2.8	14.56	−0.28
Water utilities	0.3	−1.3	0.30	0.00
Construction	0.1	−3.0	0.01	0.00
Wood products	7.1	−2.8	6.94	−0.13
Nonmetallic mineral products	11.4	−2.7	9.89	−0.16
Primary metals	43.8	−3.8	13.47	−0.24
Fabricated metal products	43.2	−2.7	11.51	−0.19
Machinery and misc. manufacturing	528.5	−1.7	34.35	−0.27
Motor vehicles	136.2	−1.8	20.74	−0.23
Food and beverage	89.9	−2.7	10.13	−0.23
Textile, apparel, leather	18.4	−1.1	19.08	−0.09
Paper and printing	32.1	−2.6	12.47	−0.23
Chemicals, plastics, and rubber	246.6	−3.3	22.03	−0.38
Trade	209.0	−2.6	7.91	−0.16
Air transportation	57.4	−3.2	32.27	−0.47
Railroad transportation	15.8	−1.4	14.17	0.01
Water transportation	10.5	−3.5	18.08	−0.36
Truck transportation	41.1	−3.9	13.01	−0.32
Transit and ground passenger transportation	0.0		0.00	0.00
Other transportation and warehousing	29.8	−2.9	9.17	−0.18
Communication and information	99.2	−1.3	7.74	−0.09
Services	411.0	−1.5	3.82	−0.05
Real estate and owner-occupied housing	4.4	−1.4	0.17	0.00

[a]In billions of 2013 dollars.
[b]*Export intensity* is defined as total exports of a good divided by the total domestic production of that good.

Overall, the trade impacts of a carbon tax are modest. The policy causes slight reductions in exports and slight increases in imports. To bring about trade balance, the dollar exchange rate falls, which reduces the adverse impact on exports and lowers imports.

Tax Interactions, Revenue Recycling, and the Double Dividend

A major theme of this book is that the impacts of climate change policies depend importantly on interactions with the fiscal system. The simulation results just described for our central case carbon tax attest to the importance of these interactions. They show, for example, that the GDP and welfare costs vary substantially depending on how the revenues are recycled. The costs are largest under lump-sum rebating of the net revenues. Costs do not disappear under other forms of recycling, but they nearly vanish when revenues are recycled through cuts in the corporate income tax, which according to our model is the most distortionary.

Here we explore further the importance of fiscal interactions, considering their implications under a range of scenarios that differ according to the stringency of the carbon tax and the nature of revenue recycling. To consider different stringencies, we employ carbon tax profiles in which the initial tax rate (after the phase-in) differs while the rate of increase in the tax remains the same.

Figure 5.6 displays how the welfare costs per avoided ton vary with stringency, measured as the percentage reduction in present discounted emissions over time, for our four revenue-recycling options. Welfare costs increase with stringency, but rise less than in proportion to the magnitude of the tax. For example, with lump-sum rebates, doubling the initial tax rate from its central case value of $20 increases the welfare costs by about 44 percent—from $42 to $61. This is in keeping with economic theory. Putting aside its (important) environmental benefits, one can view the carbon tax as a supplement to preexisting distortionary taxes on factors, goods, and services. As a result, when one compares outcomes under a higher and lower carbon tax, the relative increase in the overall taxation of these goods is less than the ratio of the two carbon tax rates.[25]

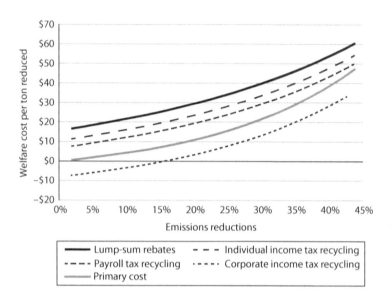

FIGURE 5.6

Welfare Costs of a Carbon Tax under Alternative Recycling Methods

The cost curves in figure 5.6 allow one to identify the tax-interaction and revenue-recycling effects. Primary cost is defined as the cost of abatement, apart from tax-interaction and revenue-recycling effects.[26] In policies with lump-sum rebating of the revenues, there is no revenue-recycling effect, so the cost in this case (shown by the dark black line in the figure) incorporates both primary cost and the cost from the tax interaction effect. The vertical distance between the dark black line and the solid gray primary cost line is the tax-interaction effect.

In the figure, the first unit of emissions abatement costs about $16 under the carbon tax with lump-sum recycling. As discussed in chapter 2, this cost of initial abatement is a critical value for environmental benefits. Our results indicate that if the environmental benefits were less than $16, no carbon tax policy with lump-sum rebates could provide net benefits. Later, we show that the benefits of initial abatement in fact exceed $16 under a range of estimates for the SCC, indicating that the tax-interaction effect does not prevent net benefits for initial abatement.

Under recycling via payroll tax cuts, individual income tax cuts, and corporate income tax cuts, the revenue-recycling effect applies. Payroll tax cuts reduce the cost of the carbon tax the least, and corporate income tax cuts reduce the cost the most, a reflection of the relative distortionary costs of these different taxes. For both payroll tax cuts and individual income tax cuts, the overall costs are higher than the primary costs, indicating that the tax-interaction effect is greater than the revenue-recycling effect for these forms of recycling. However, the cost of the corporate income tax cut policy is *below* primary cost, indicating that in this case the revenue-recycling effect dominates.

When the revenue-recycling effect outweighs both primary cost and the tax-interaction effect, the double dividend is achieved. In figure 5.6, it arises under corporate income tax recycling for levels of abatement up to about 15 percent (from a carbon tax with an initial tax rate of about $7 after phase-in, and rising at 4 percent). As indicated in chapter 2, one of the circumstances conducive to the double dividend is a situation where an environmental tax can reduce preexisting inefficiencies in the relative taxation of capital and labor. In our model, the marginal excess burden of the corporate income tax is considerably greater than the marginal excess burden of taxes on labor (see the discussion in chapter 4). This implies that on efficiency grounds, capital is overtaxed relative to labor. A carbon tax with recycling through a cut in the corporate income tax ends up reducing the tax on capital relative to labor, which improves (on efficiency grounds[27]) the relative taxation of capital and labor and accounts for the achievement of the double dividend at low levels of abatement. At higher levels of abatement, primary costs rise enough to make overall costs (the lightly dashed line) positive, despite the fact the cut in corporate tax rates causes overall cost to be less than primary cost. Thus the double dividend is not achieved at high levels of abatement.

The size of the tax-interaction and revenue-recycling effects varies with the size of preexisting tax distortions. Figures 5.7a and b display the cost curves for the lump-sum rebate and individual income tax cut policies, in scenarios where the preexisting tax rates on individual and corporate income are 25 percent lower and higher than in the central case.

In the three cases shown in figure 5.7a, there is no revenue-recycling effect since net revenues are recycled through lump-sum rebates. Because

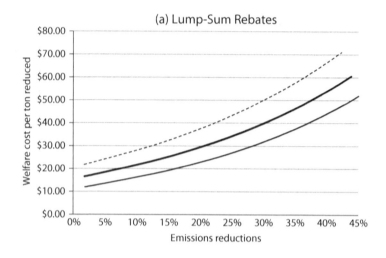

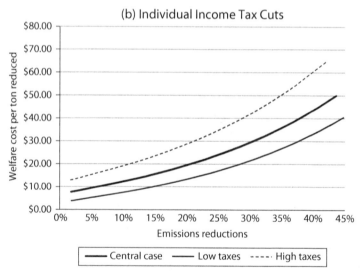

FIGURE 5.7

Preexisting Tax Rates and the Welfare Costs of a Carbon Tax

the revenue-recycling effect is absent, the differences in welfare costs in the figure represent the impact of preexisting taxes on the tax-interaction effect. Consistent with the theoretical findings offered in chapter 2, higher preexisting taxes amplify the tax-interaction effect and thereby imply higher welfare costs. The figure indicates that the size of the tax-interaction effect at the first unit of abatement is about $12 and $22 under lower and higher preexisting taxes, respectively.

The results in figure 5.7b reflect the influence of the revenue-recycling effect along with that of the tax-interaction effect. The impact of preexisting taxes on the revenue-recycling effect depends on two countervailing forces. On the one hand, higher preexisting taxes imply larger marginal excess burdens for these taxes; thus, the benefit from reducing those taxes is increasing in the size of the taxes, all else equal. On the other hand, the tax-base effect is also increasing in the size of preexisting taxes: a given carbon tax will raise less net revenue when preexisting taxes are higher. We find that this difference declines slightly with increases in preexisting taxes. Hence the latter effect dominates. Thus, higher preexisting taxes imply higher welfare costs, both by expanding the (adverse) tax-interaction effect and by diminishing the (beneficial) revenue-recycling effect. Hence, as indicated in figure 5.7b, higher preexisting taxes imply higher welfare costs, even when the revenue-recycling effect applies.

Figure 5.8 considers an alternative model specification for the degree of tax-favored consumption. As Parry and Williams (2010) and others have indicated, the presence of tax-favored consumption can improve the prospects for the double dividend from a carbon tax, particularly when the revenues are recycled through cuts in individual income tax rates. The reason is that reducing the income tax rate lowers the value of the tax deductions on favored consumption goods, which can contribute to an efficiency improvement. If this efficiency improvement is large enough, the double dividend arises.

To gauge the importance of tax-favored consumption, we consider the impact of a carbon tax in a counterfactual case where preexisting tax preferences toward consumption spending are absent. The results in figure 5.8 indicate that the absence of tax-favored consumption increases only slightly the costs of the carbon tax; that is, the *presence* of tax-favored consumption contributes only to a modest strengthening of the

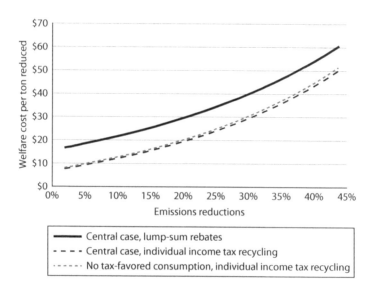

FIGURE 5.8

Tax-Favored Consumption and the Welfare Costs of a Carbon Tax

revenue-recycling effect from personal income tax cuts, relative to the case where the treatment is absent.[28] Other things equal, the impact of tax-favored consumption would have to be considerably stronger for the double dividend to arise.[29]

Climate and Nonclimate Benefits

To monetize the value of the emissions reductions, we utilize the U.S. government's estimates of the social cost of carbon (SCC).[30] We briefly referred to the SCC in our discussion earlier of the theoretically optimal carbon tax rate. The SCC is the monetary value of damages, from time t forward, attributable to the release of an additional ton of carbon dioxide into the atmosphere at time t. Correspondingly, the SCC indicates the *benefit* from the *reductions* in CO_2 emissions induced by climate policy.

The SCC is highly uncertain, and the U.S. government reports a wide distribution for the SCC estimates. In our initial comparisons of climate benefits with the non-climate economic costs, we utilize the most widely cited estimate of the SCC time-profile—the one that results when future climate damages are discounted using a 3 percent discount rate. In this case, the SCC in 2017—that is, the estimated damage from a single metric ton of CO_2 released in 2017—is $42.69 (in 2013 dollars). The SCC increases at a rate of 1–2 percent per year, so that the estimated damage from a single metric ton of CO_2 released in 2050 is $77.52 (in 2013 dollars). We multiply these central case estimates of the SCC by the policy-induced reductions emissions in each period.[31]

Table 5.10 displays the present value of our climate benefit estimates (using the reference case discount rate). In all cases, the monetized value of climate benefits greatly exceeds the welfare costs of the carbon tax policy. The excess of climate benefits over welfare costs ranges from $2.3 trillion to $4.4 trillion, depending on the nature of revenue recycling. In the next section, we show that this result holds across a range of carbon taxes and estimates for the SCC.

Table 5.10 also includes two types of nonclimate external benefits. One is an energy security benefit that relates to avoided dependence on imported oil. Brown and Huntington (2013) estimate an *energy security premium* of $4.99 per barrel of imported oil (in 2010 dollars). This is their estimate of the expected value of the additional macroeconomic costs relating to increased import dependence. In table 5.10, the row with the label "energy security premium" is based on this estimate and refers to the gain from reducing oil imports. We convert the Brown-Huntington estimate to 2013 dollars, employing a value of $5.33. Under most recycling methods, this value implies a benefit of about $79 billion. Under recycling through corporate income tax cuts, the energy security benefit increases to $105 billion, reflecting increased domestic production of crude oil and its backstop substitute.

The other nonclimate benefit reflects avoided health costs connected with the reduction of local air pollutants. The combustion of fossil fuels gives rise to emissions of a number of local air pollutants, including carbon monoxide (CO), nitrogen oxide (NO_x), sulfur dioxide (SO_2), volatile organic compounds (VOCs), ammonia (NH_3), and particulate matter (PM_{10} and $PM_{2.5}$).[32] Each of these pollutants—termed *criteria air pollutants* under the Clean Air Act—is harmful to human health. To measure the impact of climate policies on the national emissions levels of these

TABLE 5.10 Climate and Nonclimate Benefits of a Carbon Tax
Under Alternative Recycling Methods

	RECYCLING METHOD			
	LUMP-SUM REBATES	CUTS IN EMPLOYEE PAYROLL TAXES	CUTS IN INDIVIDUAL INCOME TAXES	CUTS IN CORPORATE INCOME TAXES
Welfare costs[a]	3,356	2,845	2,480	1,069
Climate benefits[b]				
–Average, 3 percent discount rate	5,679	5,664	5,646	5,454
Co-Benefits				
–Energy security premium[c]	79	79	80	105
–Local pollution benefits[d]	8,455	8,241	8,092	7,033
–NO_x	701	692	684	618
–$PM_{2.5}$	3,337	3,143	3,009	2,072
–SO_2	4,417	4,407	4,398	4,343
Total benefits	14,213	13,983	13,818	12,592
Climate benefits net of welfare costs	2,322	2,819	3,167	4,385
Total benefits net of welfare costs	10,856	11,139	11,338	11,523

[a]Welfare costs are the negative of the equivalent variation, expressed in billions of 2013 dollars.
[b]Billions of 2013 dollars. Based on Interagency Working Group central estimate of the time profile for the social cost of carbon, which employs a 3 percent discount rate to discount future climate impacts. See text for details on the method of calculation.
[c]Billions of 2013 dollars. In present value and based on assumption of a $5.33 premium per avoided barrel of imported oil.
[d]Billions of 2013 dollars. Based on cumulative emissions reductions, valued using EPA estimates of benefits per avoided ton.

criteria pollutants, we use emissions data from the EPA's 2014 National Emissions Inventory, which reports total tons of emissions of each pollutant by 60 separate sources. We then map the emissions to benchmark E3 industrial outputs, and consumption of different intermediate inputs and consumption goods to create emissions factors—emissions per unit of fuel input. The emissions factors for a given fuel input generally differ according to the industry that uses the fuel as well as the commercial or residential consumer of the fuel.[33]

TABLE 5.11 Criteria Pollutant Emissions Reductions
—*Carbon Tax with Lump-Sum Rebates*

POLLUTANT	REFERENCE CASE LEVEL OF EMISSIONS[a]			
	2020	2030	2040	2050
CO	69.25	77.19	86.34	96.92
NO_x	14.98	16.32	17.91	19.87
PM_{10}	21.46	24.94	29.00	33.79
$PM_{2.5}$	5.48	6.35	7.37	8.56
SO_2	5.08	5.14	5.17	5.27
VOC	61.47	71.59	83.54	97.57
NH_3	4.54	5.36	6.31	7.38
	PERCENTAGE CHANGE IN EMISSIONS			
	2020	2030	2040	2050
CO	−2.2	−3.3	−4.2	−4.6
NO_x	−7.4	−10.6	−12.3	−12.7
PM_{10}	−1.6	−2.4	−2.8	−3.0
$PM_{2.5}$	−2.3	−3.4	−3.9	−4.0
SO_2	−26.8	−40.3	−47.1	−49.3
VOC	−1.1	−1.9	−2.5	−2.9
NH_3	−1.1	−1.8	−2.4	−2.7

[a]Million metric tons (except for NH_3, which is in millions of short tons)

Table 5.11 displays the reference case levels for each of the seven local air pollutants, along with the percentage change in emissions in the central case carbon tax with lump-sum rebates. Reductions in coal use significantly lower SO_2 emissions in the U.S. NO_x emissions from the electric power sector are also reduced by a significant amount. The carbon tax has less pronounced impacts on emissions of VOCs and NH_3. The health benefits of these reductions depend on both the source and the location of the emissions, as well as on ambient air concentrations of these pollutants and weather patterns that disperse the pollutants across the country. Hence it is very difficult to assign dollar values to emissions reductions at the national level. In fact, the EPA does not attempt to quantify benefits per ton reduced for CO, PM_{10}, VOCs, and NH_3. However, the EPA has calculated the national average benefit per ton of emissions reductions from $PM_{2.5}$, NO_x, and SO_2 for 17 sources of emissions.[34] We map these benefit-per-ton estimates into the E3 model to calculate overall co-benefits of reduced emissions of $PM_{2.5}$, NO_x, and SO_2.

Table 5.10 reports the monetary value, in billions of 2013 dollars, of the reduced emissions of these three pollutants from the implementation of an economywide carbon tax. The overall value of the co-benefits from reduced NO_x, $PM_{2.5}$, and SO_2 emissions are $700 billion, $3,300 billion, and $4,400 billion, a total of more than $8,400 billion. Thus, even if climate benefits of reduced CO_2 were zero, a carbon tax would still produce significant net benefits through reductions in these three local air pollutants. Further, this understates the net benefits because we have not included the health benefits from reductions in CO, PM_{10}, VOCs, and NH_3 or the non-$PM_{2.5}$ benefits from reduced NO_x (a precursor to tropospheric ozone pollution) or SO_2 (a contributor to acid rain).

The nonclimate benefits far exceed the climate benefits. Thus, when nonclimate benefits are taken into account, the net benefits increase substantially. In our central case, the excess of climate and nonclimate benefits over welfare costs is more than $10,000 billion in present value, representing an increase of more than 1.3% of the present value of reference case wealth.

Figures 5.9a–d show how the net climate benefits (climate benefits minus welfare costs) vary with stringency under the four recycling methods and the four different calculations of the SCC given focus by the U.S. government's interagency task force. The previously described central estimate for net climate benefits employed values for the SCC based on a 3% discount rate. Here we obtain a range of estimates for net climate benefits by employing different time-profiles of the SCC, applying the central SCC time profile along with three other time profiles given focus by the U.S. government's Interagency Task Force. One is based on a 2.5 percent discount rate, a second is based on a 5 percent discount rate, and a third is the 95th percentile value from the distribution of SCC estimates that result with a 3 percent discount rate.[35]

When stringency is low, under each of the four benefit estimates the net climate benefits are positive, indicating that the size of the tax-interaction

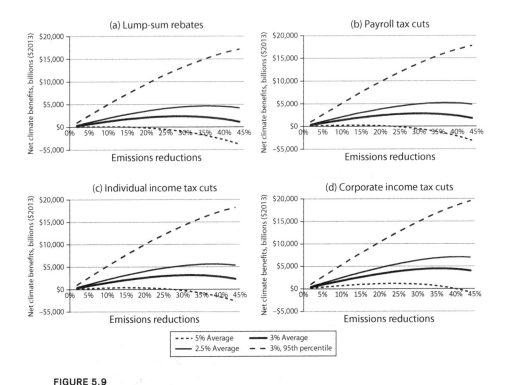

FIGURE 5.9

Net Climate Benefits by Recycling Option

effect does not lead to negative net benefits at the initial level of abatement. As stringency increases, the level of net climate benefits increase, peak, and then decrease, as marginal costs start to exceed marginal benefits. Under three of the four SCC scenarios shown, net climate benefits remain positive even at abatement levels as high as 45 percent. However, in the low SCC scenario (involving the use of a 5 percent discount rate), net benefits become negative at lower abatement levels.

The net benefit calculations in figures 5.9a–d ignore the nonclimate benefits discussed earlier. In figures 5.10a–d, the net benefits include non-climate benefits as well as climate benefits. In these figures, under all of the SCC estimates and levels of stringency considered, the total net benefits are positive—and often considerably positive.

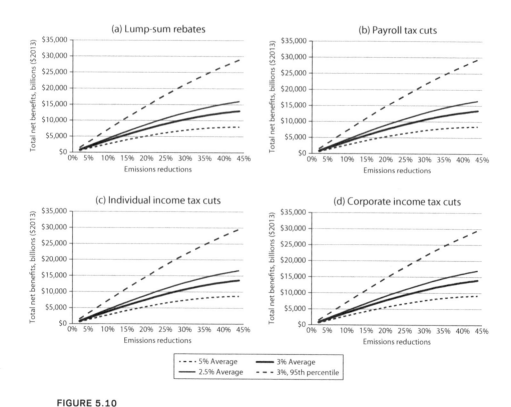

FIGURE 5.10

Total Net Benefits by Recycling Option

Alternative Parameter Specifications

Table 5.12 considers the sensitivity of the welfare costs to alternative values for key parameters.[36] The carbon tax policy here is one that achieves a 30 percent reduction in cumulative emissions. In each case, the tax is phased in over three years and the tax's growth rate is the same (four percent annually). The initial post phase-in tax (and associated height of the tax's entire time profile) will differ, however, to assure that the policy achieves the same 30-percent cumulative emissions reduction under the various parameter assumptions.[37] The welfare costs are displayed under the four recycling alternatives we have emphasized.

TABLE 5.12 Welfare Costs per Ton of Emissions Reductions
Under Alternative Parameter Assumptions

	RECYCLING METHOD			
	LUMP-SUM REBATES	CUTS IN EMPLOYEE PAYROLL TAXES	CUTS IN INDIVIDUAL INCOME TAXES	CUTS IN CORPORATE INCOME TAXES
Benchmark parameter specifications	$39.83	$33.62	$29.27	$12.11
Elasticity of labor supply (φ) —Central value = 0.3				
$\varphi = 0.1$	$32.06	$29.59	$25.03	$7.80
$\varphi = 0.5$	$46.96	$38.85	$34.92	$18.37
Intertemporal elasticity of substitution ($1/\sigma$) —Central value = 0.5				
$1/\sigma = 0.25$	$37.25	$30.54	$26.40	$11.33
$1/\sigma = 0.75$	$40.84	$34.85	$30.45	$12.41
Adjustment costs ξ —Central value = 7				
$\xi = 2$	$38.25	$32.19	$28.00	$9.86
$\xi = 14$	$41.23	$34.92	$30.39	$14.56
Generator elasticity of substitution (σ_{gen}) —Central value = 3				
$\sigma_{gen} = 2$	$45.24	$38.36	$33.56	$14.46
$\sigma_{gen} = 4$	$36.89	$31.01	$26.88	$10.69

Higher (lower) values for the uncompensated elasticity of labor supply φ imply that the preexisting distortions from labor taxes are larger (smaller). This in turn implies a larger (smaller) tax-interaction effect and larger (smaller) welfare costs. Changing this elasticity has roughly similar impacts across the four recycling scenarios.

The intertemporal elasticity of substitution in consumption ($1/\sigma$) regulates the elasticity of capital (savings) with respect to changes in the rate of interest, with a higher $1/\sigma$ implying greater elasticity. As with labor supply, more elastically supplied capital implies larger preexisting distortions, a larger tax-interaction effect, and greater welfare costs from a given carbon tax. Increasing the intertemporal elasticity has the largest impact in percentage terms under corporate income tax recycling. This squares with the fact that the elasticity affects saving behavior, and that among the recycling options, the corporate tax cut is the most closely connected to saving.

The adjustment cost parameter ξ affects the responsiveness of the demand for new capital to policy-induced changes in the return to capital, with higher values of ξ implying greater adjustment costs and less responsive demand. As with a higher-value intertemporal elasticity of substitution in consumption, lower adjustment costs imply more elastic capital. Although this yields higher absolute welfare costs from a given carbon tax, it also leads to greater emissions reductions, reflecting a decrease in the primary cost of abatement. In our simulations, the latter effect is large enough to imply lower costs per ton of emissions reduction when adjustment costs are low.

A higher elasticity of substitution between coal-fired, other-fossil, and non-fossil electricity reduces the overall costs of the carbon tax by lowering the cost of fuel switching within the power sector. The implications of a higher elasticity are of similar magnitude across the four recycling scenarios.

Summary

We've considered a range of carbon tax time profiles and a range of revenue-recycling options under each profile. In the central case, the carbon tax starts (after a three-year phase-in) at $20 per ton and increases at a rate of 4 percent per year. That carbon tax yields significant reductions in emissions of CO_2, lowering cumulative emissions by 29 percent from 2020 to 2050.

The economic costs of achieving these reductions depend importantly on the ways that revenues are recycled. We have focused on four types of revenue recycling: cuts in the rates of payroll taxes, individual income taxes, and corporate income taxes, and lump-sum rebates. These costs are significantly lower when revenues are recycled through tax rate cuts as opposed to rebates. Compared with the lump-sum rebate case, the costs are 15, 26, and 67 percent lower when recycling takes the form of rate cuts in the payroll tax, individual income taxes, and the corporate income tax, respectively (table 5.3). This ranking of the cost savings under the different recycling options is robust across a wide range of alternative parameter specifications.

In our central case, the carbon tax policy involves an aggregate cost to the economy (before accounting for the economic benefits associated with lower climate-related damages): the double dividend does not arise. However, the double dividend is obtained when the carbon tax time profile is less stringent than in our central case and recycling is via cuts in the corporate income tax. In particular, if the initial tax rate (after a three-year phase-in) is no more than $7, then a carbon tax that rises at 4 percent per year combined with corporate tax recycling yields the double dividend. Although our model suggests that a carbon tax with an initial rate of $7 (after a three-year phase-in) and rising 4 percent would be a zero-cost option, the model indicates it would reduce emissions by a much smaller amount than what the Obama administration had pledged under the Paris agreement. It would lead to CO_2 emissions in 2025 that are about 19 percent below 2005 levels, while the Obama administration pledge called for a reduction of 26–28 percent.[38]

When economic benefits from avoided climate damages are valued using a central estimate of the social cost of carbon, the carbon tax's climate-related benefits exceed the its gross costs over a wide range of abatement levels. The carbon tax also generates significant reductions in local air pollutants, and the health benefits from these reductions are larger than the policy's climate-related benefits. Accounting for these nonclimate benefits significantly amplifies the carbon tax's benefits and implies that its net benefits are positive over a wide range of abatement levels even when lower values of the SCC are assumed.

CAP AND TRADE

Cap and trade, the other key form of CO_2 emissions pricing, influences the economy in much the same way that a carbon tax does. Whether imposed upstream, midstream, or downstream, it induces reductions in the use of carbon-based fuels and their products all along the supply chain. The main difference is that under a carbon tax the regulator sets the price of emissions (this is the tax rate) while under cap and trade the regulator sets the allowable quantity of emissions.

Modeling Cap and Trade

In the model, cap and trade involves the introduction of an exogenous number of allowances A_t in each period t. This number implies a limit in aggregate emissions from covered sources in period t. If the emissions limit is binding, there will be a positive price of allowances: $p_t^a > 0$.

The model's treatment of cap and trade parallels its treatment of the carbon tax in several ways. First, as with the carbon tax, emissions are calculated using a carbon coefficient corresponding to the tons of carbon embedded in a unit of each fossil fuel. Thus, a firm that purchases fossil fuels must hold allowances accounting for all the emissions associated with those fuels—both the emissions the firm generates and the emissions from combustion downstream of its products. This treatment parallels the point of regulation described earlier for the carbon tax.

The modeling of the per-unit price of emissions is also very similar under cap and trade to its treatment under the carbon tax. For each unit of input of fossil fuels, firms must pay the allowance price times the carbon coefficient. As under the carbon tax, the price paid for domestic fuel i by industry j is $p_{ij}^d = p_i(1+\tau_{ij}^I)+d_j c_{ij} p^a$. Here we distinguish between the carbon tax and cap and trade by using p^a to denote the price of an emissions allowance. An analogous equation holds for foreign fuels, and again it is the same as under the carbon tax policy.

The positive price applies at the margin, even when allowances are freely allocated rather than auctioned out.[39] If, on one hand, the firm wishes to emit more than the amount authorized by its free allowances, then each additional unit of emissions requires further purchases at the price p^a. If, on the other hand, the firm wishes to sell some of its allowances, each unit of emissions will entail a cost by reducing the number of allowances it could sell. Thus, regardless of whether firms receive free allowances or must purchase them through an auction, at the margin, firms face a price for emitting, and they will minimize costs by reducing emissions until the marginal cost of reducing emissions equals the market price for allowances. The positive price at the margin suggests similar producer decisions between policies that involve free allowance allocation and those involving auctioned allowances.

Still, as discussed in chapter 2, free allowances and auctioning have different implications for firms' profits. Firms benefit by receiving free allowances rather than having to purchase them. Free allowances have a value that is inframarginal—this value is a lump-sum gain. Accordingly, the value of allowances received for free is modeled as an increase in LS in equation (B.16) of appendix B.

In terms of its impact on input decisions and investment, the emissions price from cap and trade functions as a carbon tax. As with the carbon tax, by differently affecting the prices of various energy goods, the emissions price from cap and trade causes cost-minimizing firms to change the mix of their energy inputs.

In addition, to the extent that it raises the prices of carbon-intensive sources of energy, cap and trade induces firms to substitute toward less carbon-intensive sources of energy and toward other material inputs or labor. Higher-priced carbon-based energy also changes the relative costs of production across industries, altering the relative demands for the outputs of different industries. Cap and trade also affects the prices of capital goods, thereby influencing the composition and levels of investment by various industries. All these responses parallel those generated by a carbon tax.

Central Case Specifications and Results

Many of the important design dimensions of a cap-and-trade policy match those of a carbon tax. They include the time profile of the aggregate cap, the points of regulation (i.e., where in the supply chain allowances must be offered to justify emissions), and the breadth of industry coverage. A further design decision is whether to give allowances free or introduce them via an auction. In addition, to the extent that allowances are auctioned, key design decisions are the extent to which the policy-generated revenues will be recycled and the form that this recycling will take. We explore these dimensions later.

To facilitate comparisons with the carbon tax, in our central case for cap-and-trade policy we impose a time profile of emissions limits $Z_t(t=1,2,...)$ that matches the emissions time profile that resulted under our central case carbon tax when recycling involved lump-sum rebates. We also impose the same point of regulation and coverage as for the central case carbon tax.

Table 5.13 displays the GDP and welfare costs of the cap-and-trade program with 100 percent auctioning under our four main revenue-recycling options. It also shows these costs for a policy with 100 percent free allocation. The economic costs of a cap-and-trade policy with 100 percent auctioning and lump-sum rebating of revenues are identical to those seen previously for the carbon tax with lump-sum recycling. Given that firms face the same marginal requirements under this cap-and-trade system as under the carbon tax, and since revenues are recycled the same way as well, cap and trade yields the same emissions price time profile as the one for the carbon tax, and the economic responses are also the same.[40]

When the revenues from the auction are recycled through cuts in marginal tax rates, the economic costs are slightly higher than those shown in table 5.3 for the carbon tax with the same marginal rate cuts. In all the simulations recorded in table 5.13, we match the emissions time profile Z_t to the time profile from the carbon tax policy with lump-sum recycling. In every period, the aggregate cap matches the aggregate emissions for that period under the carbon tax in the lump-sum recycling case, but not in the other carbon tax cases. Compared with the lump-sum case, in the other recycling cases the economic boost from tax rate cuts leads to higher demand for allowances and, given the fixed total number of allowances, a slightly higher allowance price. This underlies the slight differences between the costs of cap and trade and the carbon tax in the tax cut recycling cases.[41]

The far-right-hand column of table 5.13 considers the case of 100 percent free allocation. Such allocation involves higher costs than cap and trade with auctioning. The GDP costs are about 8 percent higher than the cost under the most costly policy that involves 100 percent auctioning—the policy with 100 percent auctioning and lump-sum rebates. They are more than 200 percent higher than the costs in the least costly auctioning case—the case of 100 percent auctioning with recycling through corporate income tax cuts.

It is easy to see why free allocation involves higher costs than the auctioning cases where revenues are devoted to cuts in preexisting distortionary taxes. In those cases, auctioning enjoys a beneficial revenue-recycling

TABLE 5.13 Implications of Alternative Allocation and Recycling Methods for GDP and Welfare Costs of Cap and Trade

	100 PERCENT AUCTIONING				100 PERCENT FREE ALLOCATION
	LUMP-SUM REBATES	CUTS IN EMPLOYEE PAYROLL TAXES	CUTS IN INDIVIDUAL INCOME TAXES	CUTS IN CORPORATE INCOME TAXES	INCREASES IN INDIVIDUAL INCOME TAXES
GDP costs[a]					
—as percent of reference GDP	0.64	0.49	0.45	0.21	0.69
—per ton of CO_2 reduced[b]	$98.02	$76.37	$69.74	$31.97	$106.54
Welfare costs[c]					
—as percent of wealth	$3,356.47	$2,860.78	$2,517.14	$1,234.10	$3,610.98
	0.41	0.35	0.31	0.15	0.44
—per $1 of gross revenue	$0.43	$0.36	$0.32	$0.14	n/a
—per ton of CO_2 reduced	$42.12	$35.90	$31.59	$15.49	$45.32

[a]GDP costs measured as present value of real GDP loss, 2017–2050, using a 3 percent real interest rate.

[b]Present value of cumulative tons reduced, using a 3 percent real interest rate.

[c]Welfare costs are the negative of the equivalent variation, expressed in billions of 2013 dollars.

effect. What is less obvious is why free allocation involves higher costs than the case of auctioning with lump-sum recycling. By construction, the emissions allowance prices in these two cases are the same. Both cases also produce similar impacts on the tax base. However, the revenue effects are different. In the case of free allocation, the negative tax-base effect compels the government to raise preexisting tax rates to achieve revenue neutrality. In contrast, in the auctioning case, the gross revenues from the auction can be used to offset the adverse tax-base effect. Thus, the rates of preexisting taxes are higher in the free allocation case than in the auctioning case. This underlies the higher economic costs in the former case.[42]

Table 5.13 does not show climate and nonclimate benefits for cap and trade; they are nearly identical to the benefits of the carbon tax. In all cases, the nonclimate benefits exceed the climate benefits, and the climate benefits themselves exceed the welfare costs. Thus, even under 100 percent free allocation, the benefits of the cap-and-trade policy greatly exceed the costs.

Because free allowances do not affect producers' decisions at the margin, the way that they are allocated across industries or firms has relatively little impact on the overall economic cost of a cap-and-trade program. However, their particular distribution across industries affects the distribution of profits, since free allowances are quite valuable to firms. To highlight these differences, we consider two distributions of allowances under 100 percent free allocation. Profit impacts under the two distributions are shown in table 5.14. The table also indicates, for comparison purposes, the impacts under 100 percent auctioning.

The first distribution policy allocates allowances to what we term *primary points of regulation*—the nine industries responsible for 99.4 percent of emissions in the reference case.[43] The allowances are allocated in proportion to the 2013 emissions attributed to each of these points of regulation. When allowances are distributed this way, these industries receive very large windfall profits. As indicated in the theoretical model of chapter 2, the limited supply of allowances leads to significant increases in the prices of output for firms receiving these allowances. When firms can shift onto consumers a significant fraction of the added costs from the cap-and-trade policy, the rents they enjoy from the free allowances exceed their costs of abatement, and profits rise.

TABLE 5.14 Profit Impacts Under Cap and Trade

Percentage Changes in Present Value of Profits over the Infinite Horizon
Figures in Parentheses Are Percent of Total Allowances Allocated to Firms

INDUSTRY	100 PERCENT AUCTIONING (LUMP-SUM REBATES)	100 PERCENT FREE ALLOCATION			
		ALLOWANCES ALLOCATED TO INDUSTRIES IN PROPORTION TO 2013 EMISSIONS[a]		ALLOWANCES ALLOCATED TO TEN SELECT INDUSTRIES IN PROPORTION TO PROFIT LOSSES UNDER 100% AUCTIONING[b]	
Oil extraction	−0.1	7.0	(1.7)	−0.1	
Natural gas extraction	−23.5	−21.6	(0.7)	78.8	(39.0)
Coal mining	−45.9	−45.9		153.9	(10.0)
Electric transmission and distribution	−7.9	−7.9		26.3	(9.0)
Coal-fired electricity generation	−74.7	391.9	(28.4)	250.4	(19.8)
Other-fossil electricity generation	−18.5	136.9	(8.0)	62.2	(4.2)
Nonfossil electricity generation	62.7	62.7		62.7	
Natural gas distribution	−8.4	58.1	(8.0)	28.1	(4.4)
Petroleum refining	−6.3	387.2	(43.7)	21.1	(3.0)
Pipeline transportation	−7.2	3.1	(1.0)	24.3	(3.0)
Mining support activities	−5.5	−5.5		18.3	(4.0)
Other mining	−3.2	−3.2		−3.2	
Farms, forestry, fishing	−1.8	−1.8		−1.8	
Water utilities	−1.0	−1.0		−1.0	
Construction	−2.3	−2.3		−2.3	
Wood products	−2.0	−2.0		−2.0	
Nonmetallic mineral products	−2.3	−2.3		−2.3	
Primary metals	−3.3	13.2	(1.9)	−3.3	
Fabricated metal products	−2.1	−2.1		−2.1	
Machinery and misc. manufacturing	−1.9	−1.9		−1.9	
Motor vehicles	−1.6	−1.6		−1.6	
Food and beverage	−1.6	−1.6		−1.6	
Textile, apparel, leather	−1.7	−1.7		−1.7	
Paper and printing	−1.8	−1.8		−1.8	
Chemicals, plastics, and rubber	−2.7	2.6	(6.1)	−2.7	
Trade	−1.6	−1.6		−1.6	
Air transportation	−2.8	−2.8		−2.8	

TABLE 5.14 *Continued*

INDUSTRY	100 PERCENT AUCTIONING (LUMP-SUM REBATES)	100 PERCENT FREE ALLOCATION	
		ALLOWANCES ALLOCATED TO INDUSTRIES IN PROPORTION TO 2013 EMISSIONS[a]	ALLOWANCES ALLOCATED TO TEN SELECT INDUSTRIES IN PROPORTION TO PROFIT LOSSES UNDER 100% AUCTIONING[b]
Railroad transportation	−3.6	−3.6	12.1 (3.7)
Water transportation	−2.4	−2.4	−2.4
Truck transportation	−2.0	−2.0	−2.0
Transit and ground passenger transportation	−1.2	−1.2	−1.2
Other transportation and warehousing	−1.8	−1.8	−1.8
Communication and information	−1.1	−1.1	−1.1
Services	−1.2	−1.2	−1.2
Real estate and owner-occupied housing	−1.1	−1.1	−1.1
All industries[c]	−1.3	−0.8	−1.1

[a]Under a modified upstream cap-and-trade program, only industries that directly purchase fossil fuels are required to hold allowances. It is restricted to nine industries responsible for 99.4 percent of allowances.
[b]Gas extraction, coal mining, electric transmission and distribution, coal-fired electricity, other-fossil electricity, natural gas distribution, petroleum refining, pipeline transportation, mining support activities, railroad transportation.
[c]Weighted average, using 2013 output as weights.

The energy-related industries that are not primary points of regulation suffer larger profit losses under this free allocation policy than under the first 100-percent free allocation policy. This reflects the increase in individual income tax rates in the free allocation case.

The second distribution allocates allowances in a manner that causes the profit impacts to be more evenly distributed across industries. Here the industries receiving free allowances are the ten that suffer the highest percentage profit loss under 100 percent auctioning, with the allowances distributed in proportion to their profit loss under lump-sum rebates. The industries that would experience the largest percentage losses include several industries that are not points of regulation; that is, firms in these industries do not need to submit emissions allowances. For example, in our policy simulations, the electric transmission and distribution industry is not a point of regulation, but nevertheless can experience significant profit losses because of the reduction in demand for retail electricity driven by the cap-and-trade program. Offering firms in these industries free allowances is a way to compensate them, since these firms can sell their allowances to firms in other industries. As shown in table 5.14, all ten industries that receive allowances under this scheme enjoy windfall profits: they are overcompensated through the receipt of free allowances.

These results suggest that profits in these industries could be preserved by freely allocating less than 100 percent of the allowances and auctioning the rest. In chapter 7, we consider more limited free allocation and look for the mix of free allocation and auctioning that preserves industry profits.

Results Under Alternative Specifications

Here we consider two alternative cap-and-trade policy designs: cap and trade with intertemporal banking and borrowing, and cap and trade that applies only to the electricity sector.

Cap and Trade with Intertemporal Banking and Borrowing

Provisions for banking and borrowing give firms greater flexibility in methods of compliance, and this has the potential to reduce policy costs. Proponents of intertemporal banking succeeded in making it a feature of the Waxman-Markey energy and climate bill that passed the U.S. House of Representatives in 2009. With intertemporal banking and borrowing, firms can hold more allowances than they need to submit in the current period and "bank" them for use in future periods. They can also "borrow" allowances by meeting current obligations either by using future-period allowances that have been promised to them for free or by purchasing future-period allowances.

Kling and Rubin (1997) and Rubin and Leiby (2013) and others have shown that the cost-effectiveness of a cap-and-trade system to achieve a cumulative emissions target is enhanced through provisions for allowance banking and borrowing, provisions that offer intertemporal flexibility in the trading of allowances. In the absence of intertemporal banking and borrowing, a cap-and-trade system can help bring about equality in marginal abatement costs across firms at given times. Provisions for intertemporal banking and borrowing contribute further to cost-effectiveness by promoting equality in the present value of marginal abatement costs across time. Thus, with banking and borrowing, a given target for cumulative reductions over some time interval can be met at lower cost through changes in the timing of emissions abatement.

Here we report results from simulations allowing for intertemporal banking and borrowing of allowances over the interval 2017–2050. We require firms to have a zero bank balance (i.e., use up all banked allowances or redeem all borrowed ones) by 2050. Hence the cumulative emissions over this interval must match the cumulative emissions in the cases without banking or borrowing.

Figure 5.11a shows the emissions price time profile under cap and trade in the absence and presence of intertemporal trading. In the model, the real rate of interest is generally around 3 percent. Under our central case cap-and-trade policy (without intertemporal trading), emissions allowance prices rise faster than this rate: from 2019 (the end of the phase-in period) to 2050, these prices rise at an average rate of about 4 percent.[44] When allowance prices rise at a rate faster than the interest rate, marginal abatement costs rise at this faster rate as well, since the marginal abatement costs of cost-minimizing firms equal the allowance price. In the absence of banking and borrowing, the present value of future abatement (according to the 3 percent interest rate) is higher than the cost of current abatement. Current abatement is relatively inexpensive. This means that firms could lower their overall abatement costs if they could shift abatement from the future to the present. They can do this when they are allowed to bank allowances, that is, to hold extra allowances in early periods, bank them, and redeem them later. In the figure 5.9b simulation in which banking is allowed, firms do this. Thus, compared with our central case without intertemporal trading, emissions reductions are larger in the short term (in keeping with the nearer-term banking) and smaller in the long term (in keeping with the later redemptions).

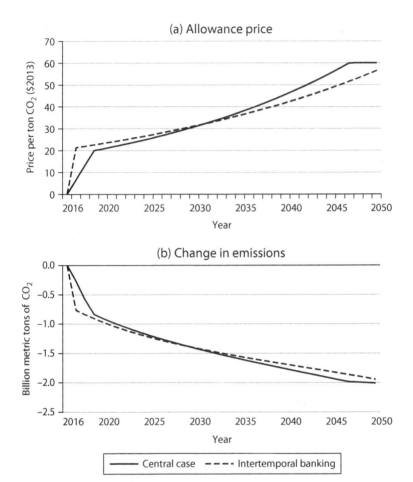

FIGURE 5.11

Cap and Trade with and without Intertemporal Banking: Allowance Prices and Changes in Emissions

Although intertemporal banking and borrowing lowers cost by equating the discounted marginal abatement costs over time, the E3 simulation results suggest the cost savings of intertemporal banking and borrowing are minimal. The central case cap and trade program with intertemporal banking and borrowing is only 1.2 percent less costly than the equivalent policy without banking and borrowing with an allowance price growth of 4 percent. We also compared cap-and-trade programs with different allowance price growth rates, with and without banking and borrowing. For growth rates of 2 percent and 6 percent, banking and borrowing reduces welfare costs by 0.2 percent and 2.2 percent, respectively. As the potential cost savings from banking and borrowing increase, the larger the discrepancy between the rate of interest and the rate of growth of allowance prices in the absence of banking and borrowing.

In sum, banking and borrowing can reduce the cost of meeting a cumulative (and finite) emissions target by optimizing the trajectory of the carbon price over time. However, under the conditions assumed in our central case cap-and-trade policy, we find the benefits of such a policy to be limited. Banking and borrowing reduces the welfare cost of our central case policy by only about 1 percent.

A Narrow (Electricity-Sector-Only) Cap-and-Trade Program

So far we have considered only broad-based emissions pricing policies. Economists generally push for broader systems on the grounds that such systems have greater potential to harvest the economy's "low-hanging fruit," that is, its low-cost abatement opportunities. Among the cap-and-trade systems in the U.S., only California's has broad sectoral coverage, as it applies to CO_2 emissions from the power sector, the transportation fuels (refining) sector, and various large industrial emitters of CO_2. The Regional Greenhouse Gas Initiative, with nine states in the northeast, only caps emissions from the electricity sector. The Clean Power Plan introduced by the Obama administration gave states the option to reduce emissions through a cap-and-trade program, but it too applied only to the power sector.

To what extent is narrower sectoral coverage a handicap in terms of efficiency? Here we address this question by comparing a national electricity-sector-only program with the national economywide program considered previously. Under the narrower policy, the firms that must submit allowances are restricted to electricity generators with carbon emissions (coal-fired generators and other-fossil generators).

We compare the narrow and broad cap-and-trade programs under alternative assumptions about the time profile of the aggregate emissions cap. In one scenario, we set the cap to approximate the time profile of power sector emissions projected by the U.S. EPA under the Clean Power Plan (CPP). States electing to institute a cap-and-trade program had the option of linking their program with cap-and-trade programs in other states and including new sources of emissions in the cap-and-trade program. Our scenario thus represents one possible outcome of the Clean Power Plan, namely, one in which all states choose to meet their obligations through cap and trade, all states link their cap-and-trade systems, and the emissions cap applies to new coal-fired and other-fossil generators. The Trump administration has indicated its intention to repeal the Clean Power Plan. Even if it is not repealed, the Trump administration is unlikely to enforce it. Still, the plan's projected level of reductions serves as a useful benchmark for power-sector-only emissions policies. We refer to this case as the *CPP stringency case.*

Figure 5.12a displays the changes in economywide emissions in the CPP stringency case and two other power-sector-only cap-and-trade systems of greater stringency. In the CPP stringency case, the emissions reductions are significantly smaller in magnitude than those under the central case cap-and-trade system examined earlier. Because it is useful to consider emissions caps that achieve reductions closer to those in the earlier cap-and-trade cases, our other power-sector-only cases involve reductions that are 50 and 100 percent larger in each period than in the CPP stringency case. These additional cases are labeled the $1.5 \times$ CPP stringency and $2.0 \times$ CPP stringency cases.

Figure 5.12b displays the allowance prices necessary to deliver the required reductions in power sector emissions under each of the three power-sector-only cap-and-trade policies. If all states joined a nationally linked cap-and-trade system, the emissions prices necessary to achieve the national CPP emissions targets are relatively modest, peaking at about $10 in 2030. With stronger emissions reduction targets, the $1.5 \times$ CPP stringency and $2.0 \times$ CPP stringency policies require higher allowance prices. The $2 \times$ CPP stringency policy requires a price more than 5 times greater than the price in the CPP stringency case, reflecting increasing abatement costs in the power sector.

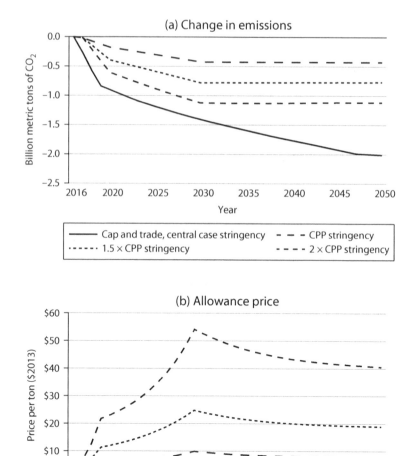

FIGURE 5.12

Broader and Narrower Cap-and-Trade Policies: Allowances Prices and Changes in Emissions

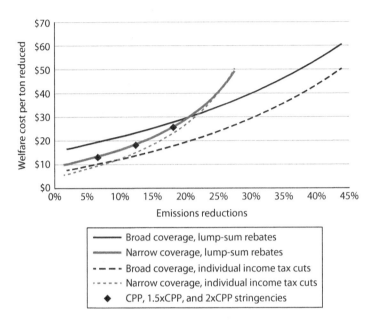

FIGURE 5.13

Welfare Costs of Broader versus Narrower Cap-and-Trade Policies

Figure 5.13 compares the costs of broad (economywide) and narrow (power-sector-only) cap-and-trade policies. By simulating each policy over a range of emissions reduction targets,[45] we produce cost curves for each policy. In all cases, we assume that emissions allowances are introduced via auction. And in every case, we consider two methods for recycling of the auction revenues: lump-sum rebates and cuts in the marginal tax rates on individual income.

The solid black line in the figure represents the welfare costs of an economywide cap-and-trade program with lump-sum rebates, and the dotted black line represents the welfare costs of the same policy with recycling through cuts in the marginal rates of individual income taxes. The gray lines represent the costs of the power-sector-only policies. The three diamonds represent the costs of the narrow policies at the three levels of stringency considered earlier.

Perhaps surprisingly, when recycling is via lump-sum rebates, the electricity-sector-only cap-and-trade program sometimes achieves reductions at *lower* cost than the broader program! Under lump-sum recycling, this is the case in all but the most stringent policies. This defies the widely supported economic axiom that broader coverage adds to cost-effectiveness.

Preexisting distortionary taxes give rise to the tax-interaction effect that is responsible for this "contrary" result. In the absence of preexisting taxes, the broader policy achieves reductions at a lower cost by equalizing firms' marginal abatement costs across sectors. However, in the presence of prior taxes, equating *firms'* marginal abatement costs across sectors does not correspond to equating, at the margin, *all* the costs associated with each sector's abatement. Rather, in this context the cost of abatement for each sector at the margin is the firm-level marginal abatement cost in that sector *plus* the tax-interaction effect associated with that sector's abatement. In the power sector, this tax-interaction cost is relatively small because the elasticity of demand for electricity is large relative to the average elasticity in other carbon-intensive sectors. The high elasticity of demand implies that when abatement takes place in the power sector, electricity prices go up relatively little and the tax-interaction effect associated with that sector's abatement is relatively small. Thus, when the power-sector-only cap-and-trade program requires the power sector to take on extra abatement relative to a broader program that requires the same overall abatement, overall costs can be lower because, at the margin, the sum of the firm-level abatement costs and the tax-interaction cost is lower in the power sector than in other sectors.

With increased stringency, the gap between the *firm-level* marginal abatement costs of the power sector and other sectors increases. (The firm-level marginal abatement costs do not account for the tax-interaction cost.) At higher levels of emissions reductions, this gap overcomes the power sector's relatively low tax-interaction costs, and the narrower policy eventually becomes more costly at higher levels of stringency.

The potential for lower costs through a narrow cap-and-trade program is much weaker, however, when recycling takes the form of cuts in marginal tax rates. In this case the high elasticity of demand for electricity lowers the potential revenue from the cap-and-trade program and weakens the (beneficial) revenue-recycling effect relative to its strength under a broader policy. This offsets the electricity sector's advantage in terms of a low tax-interaction effect. As a result, when revenue recycling takes the form of cuts in the marginal rates of individual income taxes, the narrower program is only less costly when emissions reductions are less than about 10 percent. When the narrow program is very stringent (20 percent overall emissions reductions or more), it is generally at least 50 percent more costly than the broad-based policy with identical recycling. The difference in policy costs would be even greater if revenues were used to cut corporate income tax rates.[46]

A potential liability from narrow coverage is leakage—policy-induced increases in emissions in noncovered sectors that offset reductions in the covered sector. Our power-sector-only policy causes electricity prices to rise relative to prices of goods and services in noncovered sectors. This leads to increased demands for the outputs of these other industries and causes their emissions to rise above reference case levels. Leakage may also occur as a result of the drop in demand for coal and natural gas caused by the power-sector-only policy; a drop in prices for coal and natural gas will increase demand in noncovered sectors. Leakage raises the costs of achieving given targets for economywide reductions, since caps must be tighter to compensate for increased emissions in noncovered sectors. Although leakage works toward higher costs, our results indicate that the extent of leakage and the magnitude of the leakage-related costs are relatively minor. At the low (CPP) level of stringency, we find leakage to be about 1.2 percent, meaning that about 1.2 percent of the emissions reductions in the power sector are offset through increased emissions elsewhere. At the 1.5 × CPP and 2 × CPP levels of stringency, leakage is approximately 1.4 and 1.7 percent, respectively. Accordingly, for the narrow cap-and-trade programs we consider, the additional costs attributable to leakage are not large enough to offset the cost savings connected with preexisting taxes.

Thus, limiting the scope of a cap-and-trade program to the electric power sector does not always imply a sacrifice of cost-effectiveness. Taking account of the tax-interaction effect, we find that a narrow policy can be more cost-effective if recycling is through lump-sum rebates and the required emissions reductions are modest. On the other hand, an economywide cap-and-trade program with recycling through cuts in preexisting tax rates is often more cost-effective than the equivalent narrow program.

■   ■   ■

In this chapter we have evaluated the potential impacts of two key forms of emissions pricing—a carbon tax and a cap-and-trade system. We find that judicious recycling of the revenues from a carbon tax can lower the policy's costs considerably. The costs are lowered most when revenues finance cuts in taxes on corporate income. These taxes have the largest distortionary cost in our analysis, and thus reducing the rates of these taxes offers particularly large cost savings.

For our central case carbon tax, the costs per ton are estimated to be about $42 (in 2013 dollars) when recycling is through lump-sum rebates. These costs are reduced by $11 and $28 per ton, respectively, when recycling is via cuts in individual income taxes and corporate income taxes. The relative gain from marginal rate recycling depends on the stringency of the carbon tax; it diminishes as the carbon tax becomes more stringent. When the stringency of the policy is relatively low, and recycling takes the form of cuts in the corporate income tax rate, the carbon tax policy involves negative cost: the double dividend is obtained.

The results under cap and trade often mirror those of the carbon tax. In our analysis, a cap-and-trade program with allowances introduced via an auction yields the same outcomes as a carbon tax program when the programs are of equal stringency (that is, when the cap yields allowance prices that match the carbon tax rates) and the methods of revenue recycling are the same. Auctioning the allowances and using the revenues to finance marginal tax rate cuts is particularly cost-effective. Under free allocation of allowances, there is no possibility of exploiting the revenue-recycling effect. Accordingly, under a cap-and-trade with free allocation, the costs per ton are 44 percent higher than under a program of equal stringency in which allowances are auctioned and the revenues finance cuts in individual income tax rates.

Using central estimates for the social cost of carbon, both policies yield significant net benefits: the climate-related benefits exceed the economic costs. The policies also yield net benefits if one ignores the climate benefits and only accounts for the health benefits associated with the reductions in local pollution emissions that go along with the reductions in CO_2. These results hold not just in our central case estimates but also across a range of stringencies and assumptions related to the value of climate benefits.

Both carbon taxes and cap-and-trade programs impose a price on carbon and give rise to similar incentive effects. Correspondingly, carbon taxes and cap-and-trade programs with similar designs have similar impacts on emissions and economic costs. In fact, under some policy designs our model yields identical results under the two policies. But it is worth recognizing that our model cannot consider a key difference between the two policies: under a carbon tax, there is uncertainty about the extent to which the tax will reduce emissions, while under cap and trade there is uncertainty about the emissions price (the price of allowances). Uncertainties can affect the relative costs of carbon taxes and cap and trade. Uncertainties about future allowance prices under cap and trade can have economic costs insofar as they affect general confidence and incentives to invest in low-carbon and other technologies. Uncertainties about emissions reductions from carbon taxes can add to the costs of a carbon tax to the extent that continual adjustments to the tax rate are needed to arrive at a desired emissions reduction target. In our model, in which agents have perfect foresight, these differences are not captured.

While emissions pricing has considerable support by economists, there are other ways to reduce CO_2 emissions. In chapter 6 we explore the impacts of two other approaches to emissions reductions—a federal-level clean energy standard and an increase in the federal gasoline tax—and compare their impacts with those observed in this chapter.

In this chapter we have concentrated mainly on the aggregate impacts, ignoring some distributional issues. The distribution of impacts has important implications for the fairness of climate policies as well as for political feasibility. In chapter 7 we take a further look at the policies considered in this chapter and the next. There we consider the distributional impacts across households and U.S. states and expand our analysis of distributional impacts across industries. We also investigate various compensation schemes to reduce disparities in the distribution of policy costs as well as assess the potential trade-offs between cost-effectiveness and a more even distribution of policy costs.

6

ALTERNATIVES TO EMISSIONS PRICING: A CLEAN ENERGY STANDARD AND A GASOLINE TAX INCREASE

Notwithstanding the various attractions of emissions pricing for reducing CO_2 emissions, it is useful to consider some alternative approaches. Here we focus on two: a federal-level clean energy standard (CES) and an increase in the federal gasoline tax.

Examining these alternatives is valuable for several reasons. First, although policy analysts and decision makers have displayed considerable interest in these policies, very few studies have used a single consistent framework to compare these policies with the emissions pricing policies explored in chapter 5. Second, the alternatives are similar or identical to policies with which there is already some experience at the state and regional levels. As of January 2016, twenty-nine states had implemented renewable portfolio standards (RPSs), which (as explained later) are quite similar to the CES policy focused on here. And gasoline taxes exist in all states and at the federal level. The introduction of a federal CES, or the incrementing of a federal gasoline tax, would not be a fundamental departure from existing policies.

A third reason applies to the CES policy, in particular: prior assessments have neglected a critical attraction of this policy. Economists generally view the emissions pricing policies of chapter 5 as the most cost-effective approaches for reducing CO_2 emissions, usually for the reasons articulated in that chapter. However, in making their pitch for emissions pricing, they have tended to overlook tax interactions. As we will show in this chapter, the CES has the potential to produce a smaller adverse tax-interaction effect than equivalently scaled emissions pricing policies. In some circumstances, this advantage renders the CES more cost-effective than a similarly scaled emissions pricing policy. As the analysis will indicate, much depends on the stringency and design of the CES and emissions pricing policy. In chapter 5 we observed that tax interactions could reverse conventional conclusions about policy rankings: we found that a narrower cap-and-trade program can sometimes be more cost-effective than a broader one. We now encounter another case where there is potential for such a reversal.

We start with a focus on the CES and then turn to the gasoline tax. At the end of the chapter we offer some general conclusions regarding the advantages and disadvantages of these alternatives relative to the emissions pricing policies.

A CLEAN ENERGY STANDARD

U.S. environmental policy often employs intensity standards. The descriptor *intensity* applies because these standards are defined in terms of shares or ratios. The renewable portfolio standards (RPS) mentioned earlier are examples of intensity standards. RPS policies impose a floor on the share of electricity purchased by electric utilities that comes from sources deemed renewable (e.g., electricity from wind farms or solar panels). In doing so, RPSs aim to give a boost to renewable-sourced electricity.

The CES is another example of an intensity standard. The CES also establishes a floor: in this case, it is on the ratio of "clean" electricity (electricity whose production involves relatively low emissions) to total electricity. Typically, the CES promotes a wider range of electricity sources than the RPS does by incorporating nuclear-generated electricity, which usually receives no favorable treatment (i.e., is not deemed "renewable") under an RPS. In 2012, former senator Jeff Bingaman (D-NM) sponsored the Clean Energy Standard Act of 2012, which called for a nationwide CES.[1] The policy was mentioned by President Obama in his 2012 State of the Union speech. It received widespread attention from policy analysts but had little Congressional support, as many lawmakers viewed the bill as a political liability during the run-up to the November 2012 elections.

Intensity standards also apply to the Clean Power Plan (CPP) that President Obama proposed in June 2014 to reduce emissions from existing fossil-based electric power plants. Under the CPP's default option, by 2030 each state's emissions rate—the ratio of the state's CO_2 emissions from its fossil-based power plants to the state's electricity generation from those plants—must not exceed the target rate assigned to each state.

Economists recognize some attractions of such standards but tend to offer only faint praise. On the positive side, intensity standards are seen as superior (on cost-effectiveness grounds) to some conventional policy approaches. In contrast to specific technology mandates, intensity standards give firms or facilities the flexibility to choose whatever production method meets the standard at lowest private cost. And many intensity standards (including the RPS and CES) allow credit trading, which tends to equalize marginal abatement costs across heterogeneous firms.

But economists generally view intensity standards as less cost-effective than emissions pricing policies such as emissions taxes or systems of tradable emissions allowances. The oft-cited drawback of such standards is that while they might achieve a highly efficient ratio of emissions per unit of output, they fail to promote efficient reductions in output or demand. There are two ways to meet an intensity standard defined in terms of emissions per unit of output: reduce emissions or raise output. The latter option works against the cost-effectiveness of intensity standards because it reduces firms' incentives to achieve emissions reductions through reduced output. Indeed, formal analyses indicate that an intensity standard is formally equivalent to the combination of a tax on emissions and a subsidy to output.[2] The implicit subsidy to output compromises cost-effectiveness.

In the case of the CES, the subsidy means that the price of output (electricity) will be lower than the levels that yield desired reductions in CO_2 emissions at the lowest cost. The "low" electricity prices limit the extent that emissions are reduced through the channel of lower electricity demand.

However, considerations of fiscal interactions suggest that lower electricity prices can also make a positive contribution to cost-effectiveness. As discussed in earlier chapters, the tax-interaction effect is larger to the extent that environmental policies lead to increases in the prices of output. Because it gives rise to a less pronounced increase in electricity prices, the CES leads to a smaller tax-interaction effect than emissions pricing policies that yield the same reduction in emissions from the electricity sector. Thus, while the CES's implicit subsidy to output is a disadvantage because it limits conservation, it is an advantage in terms of producing a smaller tax-interaction effect.

Which of the two opposing effects dominates? This cannot be determined a priori but depends on behavioral parameters and specifics of the policy design.[3] With the E3 model we are able to assess the costs of the CES under plausible assumptions about parameters and under a range of designs. We describe the assumptions and report the results here.

Modeling the Clean Energy Standard

The CES policy applies to electric utilities, affecting their demands for the wholesale electricity supplied by the three types of generators (coal-fired, other-fossil, and nonfossil) by imposing a floor on the ratio of "clean" electricity to total electricity. What forms of electricity are designated clean is a policy choice; we assume that the electricity from "nonfossil generators" earns this designation. In the Bingaman proposal, natural-gas-fired electricity was designated as "partially clean:" it gave partial credit for purchases of natural-gas-fired electricity. In effect, this extends the domain of "clean" and allows utilities' purchases of natural-gas-fired electricity to contribute toward their meeting the standard. In the simulations later, we consider a range of policy designs, including some that favor natural-gas-fired electricity in this way.

Let $\bar{M}_t$ denote the standard (floor) in period t. The standard implies the following constraint:

$$\left. \frac{\sum_i a_i m_i x_{it}}{\sum_i m_i x_{it}} \right] \geq \bar{M}_t \tag{6.1}$$

The left-hand side is the ratio of clean to total electricity achieved by the utility. The product $m_i x_{it}$ in the numerator and denominator is the quantity of electricity purchased in period t from generator i by the electric transmission and distribution industry, where x_{it} is the quantity of fuel i used at time t (in units of the model's data) and m_i is a scaling coefficient that converts these units to megawatt hours.[4] The symbol a_i in the numerator is an indicator variable, equal to 1 if the generator type qualifies for the standard and 0 otherwise. Partial credit to generator i is modeled by setting a_i at a value between 0 and 1. Since $\bar{M}_t > 0$, electricity from qualifying generators is subsidized while power from nonqualifying generators is taxed.

The electricity transmission and distribution industry minimizes the cost of wholesale electricity inputs subject to the constraint in equation (6.1). The price for domestically produced wholesale electricity from generator i to the transmission and distribution utility j (ignoring the time subscript) is $p_{ij}^d = p_i + p^{CES} m_i (\bar{M} - a_i)$. The "price" $p^{CES} \geq 0$ is the shadow value on the constraint in equation (6.1). As shown in the appendix B, the equilibrium value for p^{CES} will satisfy $\Sigma_i p^{CES} m_i (\bar{M} - a_i) x_i = 0$. This condition shows that the "taxes" paid for inputs of nonqualifying electricity exactly offset the "subsidies" for qualified electricity in equilibrium.

Design Dimensions

In the central case, our CES policies adopt several features of the Bingaman proposal. As in that proposal, non-fossil-generated electricity is given full credit and natural-gas-fired electricity half credit. Coal-fired generation is considered "dirty" and receives no credit.[5] We subsequently consider alternative specifications for the extent of credit for natural-gas-fired electricity.

We use the term *CES ratio* to refer to the required minimal ratio of clean generation (in megawatt hours) to total generation. We consider three policies that differ according to the required increase in the CES ratio over time. The specified time profiles for the paths of these ratios are displayed in figure 6.1. In each case, the CES begins with the business-as-usual ratio of 48 percent in 2017. The ratio rises over time, ultimately arriving at values of 70, 80, and 90 percent in 2035 and remaining at those values thereafter.

Each CES policy is compared with a carbon tax policy scaled to achieve the same cumulative emissions reductions (in present value) from the electricity sector as those from the CES policy.[6] We compare the CES policies with carbon tax policies involving various methods of recycling: lump-sum rebates, reductions in taxes on individual income (on wages, interest income, dividends, and capital gains), and reductions in taxes on corporate income.

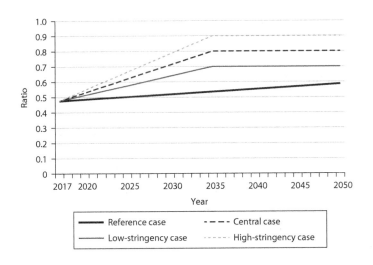

FIGURE 6.1

CES Ratio by Time and Stringency, 2017–2050

In keeping with our previous policy simulations, the CES policies considered here are revenue-neutral. As with carbon taxes, the CES policies influence the bases of existing taxes through their effects on incomes and spending. Through this channel, these policies lead to a slight reduction in gross revenue. To bring about revenue neutrality, these revenue impacts are offset through (small) increases in the marginal rates of individual income taxes.

Central Case Results

Emissions, Shadow Taxes, and Shadow Subsidies

Figure 6.2 displays the time profile of CO_2 emissions in the reference case and under the three CES policies of differing stringency. The low-, medium-, and high-stringency cases are where the CES ratio is increased to 70, 80, and 90 percent, respectively, in the long run. The kink at year 2035 reflects the fact that beginning in that year the ratios no longer increase but instead remain constant. Correspondingly, emissions no longer decline but instead increase with the growth rate of the economy.

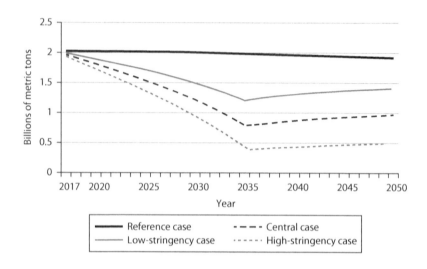

FIGURE 6.2

CO_2 Emissions from Electricity Generation, 2017–2050

As mentioned earlier the CES is equivalent to a tax and subsidy program, where a given utility purchasing electricity either faces a tax or receives a subsidy, depending on whether the electricity is produced through "clean" generation. For the electricity sector as a whole, the revenue cost of the subsidies matches the revenue yield of the taxes. Figure 6.3a through 6.3c displays the shadow tax and subsidy rates applying to the electricity from the three generators. For each generator i, these correspond to the values $p^{CES}m_i(\bar{M}-a_i)$ described earlier.

From 2017 through 2035, the shadow tax on electricity from coal-fired generators tends to rise, reflecting the increasing stringency of the CES over time and the associated need to induce greater substitution away from coal-fired electricity. Starting in 2035, the CES ratio is held constant, and the shadow tax no longer increases. Indeed, it falls. The pattern for nonfossil-generated electricity is the mirror image of the one for coal-fired electricity generation. Because nonfossil generation is deemed clean, the shadow tax is negative; that is, the utility receives a subsidy for purchasing electricity from nonfossil generation. This shadow subsidy expands over the medium term before contracting starting in 2035. In the central case simulations considered here, other-fossil generators receive

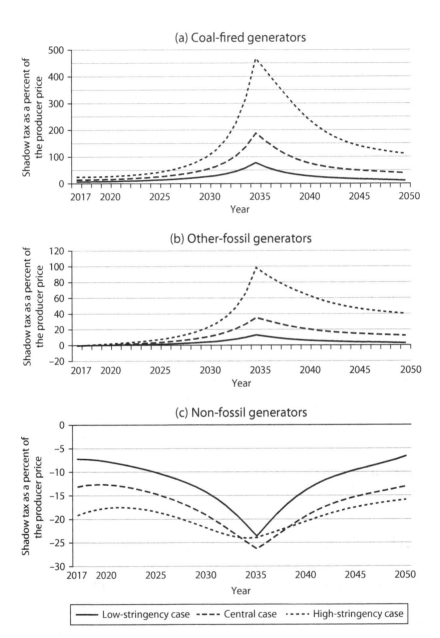

FIGURE 6.3

Shadow Tax/Subsidy on Electricity Generators

partial credit. In the initial years of the policy, the required CES ratio $\bar{M}_t$ is slightly less than the partial credit, which implies that the utility receives a subsidy for purchasing other-fossil-generated electricity. Over time, the CES ratio $\bar{M}_t$ is increased and eventually exceeds the credit. Correspondingly, the shadow subsidy becomes a shadow tax.

Impacts on Prices

As indicated earlier, the potential advantage of the CES over an emissions pricing policy is that it yields smaller increases in electricity prices and the price level and thus produces a smaller tax-interaction effect. Figures 6.4a,

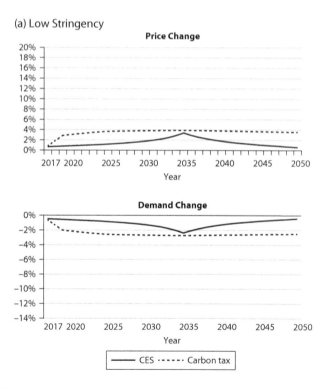

FIGURE 6.4

Percentage Change in Price of and Demand for Retail Electricity

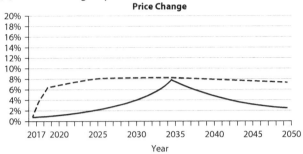

Price Change

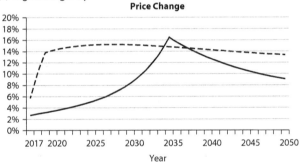

Demand Change

(c) High Stringency

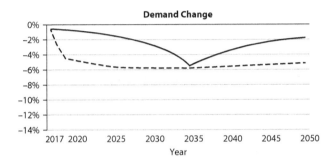

Price Change

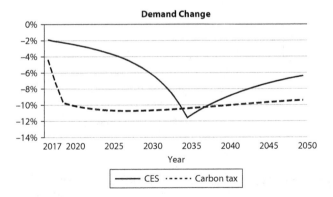

Demand Change

——— CES ┈┈┈┈ Carbon tax

FIGURE 6.4 (*Continued*)

b, and c compare the two policies in terms of their impacts on electricity prices and demands, under CES policies of different stringencies. Greater stringency implies larger increases in prices and correspondingly larger reductions in demand. Under the CES policies, the price increases (and demand reductions) are small in the short run and increase over time. After the three-year phase-in, the carbon tax policies display relatively flat changes in electricity prices and demands. In all the stringency cases, the reduction in demand for electricity is greater under the carbon tax than under the CES policy in both the short and long run, in keeping with the fact that electricity prices increase by a larger amount under the carbon tax in nearly every period. Under the CES policies, the reductions in long-run electricity demand are over five times larger in the high-stringency case than in the low-stringency case. Under the carbon tax policies, the long-run demand reductions are less than four times larger under high stringency than under low stringency. The smaller demand reduction increases under the carbon tax reflect the efficient use of demand reductions under emissions pricing.

Figures 6.5a, b, and c display, for the three CES policies of different stringencies, the effects of the CES and carbon tax policies on the price

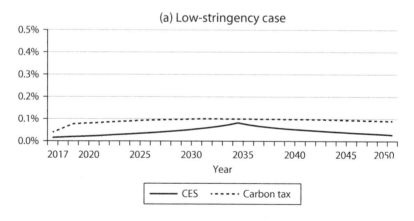

FIGURE 6.5

Percentage Change in Price of Consumption Bundle

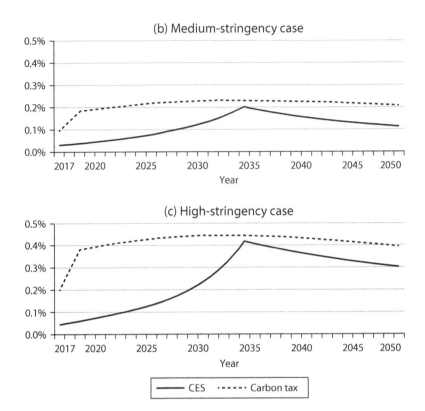

FIGURE 6.5 (*Continued*)

index for the Cobb-Douglas composite of consumption goods.[7] It shows, for the interval 2017–2050, the percentage change in this price relative to its value in the reference case. The percentage change in other aggregate prices, such as the price of the producer price index, displays a time profile and pattern similar to those of the percentage change in the price of the consumption bundle. Because emissions pricing boosts electricity prices more, electricity-only carbon taxes have a more pronounced effect on the overall price level than does an equivalent CES policy. The carbon tax's price increases are larger over the entire time interval shown. The difference in the price impact is especially large in the early years of the policy, reflecting the flat profile of the carbon taxes. In figure 6.5 the short-run deviation in the percentage increases in prices is largest for the most stringent policy, although the long-run ratio of these percentage increases declines with stringency.

Profit Impacts

Table 6.1 displays the profit impacts of the medium-stringency CES policy and the equivalent electricity-sector-only carbon tax. The medium-stringency CES policy requires a clean energy ratio of 80 percent by 2035 and is most similar to the Bingaman proposal. While the economywide carbon tax gives rise to profit losses of 1 percent or more for a large majority of the industries, the CES policy and the electricity-only carbon tax produce large profit impacts only in the electricity industries and industries (such as coal mining) with close ties to the power sector. Compared with the electricity-only carbon tax, the CES imposes larger profit losses on the fossil fuel generators and larger profit increases in nonfossil generation, in keeping with the fact that under the CES, the power sector relies primarily on fuel substitution to reduce emissions. The CES and power sector carbon tax policies produce nearly identical profit impacts on the coal mining industry.

Costs: Welfare Impacts

Table 6.2 shows the welfare impacts of the medium-stringency CES policy, the equivalent power-sector-only carbon tax, and the central case economywide carbon tax considered in chapter 5. As the former two policies are narrower and do not require emissions reductions as extensive as those under the central case carbon tax, their overall costs are much lower. The relative costs of the CES and the power-sector-only carbon tax depend on the nature of revenue recycling. The CES policy is more cost-effective than a power-sector-only carbon tax with lump-sum recycling, but about the same in cost-effectiveness as a power-sector-only carbon tax with individual income tax recycling.

TABLE 6.1 Profit Impacts Under the Clean Energy Standard
Percentage Changes in the Present Value of Profits over the Infinite Horizon

INDUSTRY	CLEAN ENERGY STANDARD	POWER-SECTOR CARBON TAX MATCHING CES EMISSIONS		CENTRAL CASE ECONOMYWIDE CARBON TAX	
	INDIVIDUAL INCOME TAX INCREASES	LUMP-SUM REBATES	INDIVIDUAL INCOME TAX CUTS	LUMP-SUM REBATES	INDIVIDUAL INCOME TAX CUTS
Oil extraction	-0.1	-0.1	-0.1	-0.1	-0.1
Natural gas extraction	-5.7	-3.6	-3.6	-23.5	-23.3
Coal mining	-25.3	-27.8	-27.8	-45.9	-45.7
Electric transmission and distribution	-1.9	-4.5	-4.4	-7.9	-7.6
Coal-fired electricity generation	-62.5	-56.6	-56.7	-74.7	-74.6
Other-fossil electricity generation	-19.0	-5.9	-5.9	-18.5	-18.3
Nonfossil electricity generation	78.0	32.0	32.2	62.7	63.4
Natural gas distribution	-0.2	-0.1	-0.1	-8.4	-8.1
Petroleum refining	-0.2	-0.2	-0.1	-6.3	-6.1
Pipeline transportation mining support	-0.2	-0.5	-0.4	-7.2	-7.0
Activities	-0.2	-0.8	-0.7	-5.5	-4.9
Other mining	-0.4	-0.7	-0.6	-3.2	-2.8
Farms, forestry, fishing	-0.1	-0.2	-0.2	-1.8	-1.6
Water utilities	-0.1	-0.2	-0.2	-1.0	-0.8
Construction	-0.1	-0.4	-0.3	-2.3	-1.8

Wood products	−0.2	−0.3	−0.3	−2.0	−1.7
Nonmetallic mineral products	−0.2	−0.4	−0.3	−2.3	−2.0
Primary metals	−0.2	−0.4	−0.3	−3.3	−3.1
Fabricated metal products	−0.2	−0.3	−0.2	−2.1	−1.8
Machinery and misc. manufacturing	−0.1	−0.2	−0.2	−1.9	−1.6
Motor vehicles	−0.1	−0.2	−0.2	−1.6	−1.3
Food and beverage	−0.1	−0.2	−0.2	−1.6	−1.4
Textile, apparel, leather	−0.2	−0.2	−0.2	−1.7	−1.4
Paper and printing	−0.1	−0.2	−0.2	−1.8	−1.6
Chemicals, plastics, and rubber	−0.2	−0.2	−0.2	−2.7	−2.4
Trade	−0.1	−0.2	−0.2	−1.6	−1.4
Air transportation	−0.1	−0.2	−0.1	−2.8	−2.6
Railroad transportation	−0.8	−1.1	−1.1	−3.6	−3.4
Water transportation	−0.2	−0.3	−0.3	−2.4	−2.2
Truck transportation	−0.1	−0.2	−0.2	−2.0	−1.8
Transit and ground passenger transportation	−0.1	−0.2	−0.2	−1.2	−1.0
Other transportation and warehousing	−0.1	−0.2	−0.2	−1.8	−1.7
Communication and information	−0.1	−0.2	−0.2	−1.1	−0.9
Services	−0.1	−0.2	−0.2	−1.2	−1.0
Real estate and owner-occupied housing	−0.1	−0.2	−0.2	−1.1	−0.9
All industries[a]	−0.1	−0.2	−0.2	−1.3	−1.1

[a]Weighted average, using 2013 output as weights.

TABLE 6.2 Welfare Costs and Benefits: Clean Energy Standard versus Carbon Tax

	CLEAN ENERGY STANDARD	POWER-SECTOR CARBON TAX MATCHING CES EMISSIONS		CENTRAL CASE ECONOMYWIDE CARBON TAX	
	INDIVIDUAL INCOME TAX INCREASES	LUMP-SUM REBATES	CUTS IN INDIVIDUAL INCOME TAX	LUMP-SUM REBATES	CUTS IN INDIVIDUAL INCOME TAX
Welfare costs	$586.35	$724.23	$590.44	$3,356.47	$2,479.64
—as percent of wealth	0.07	0.09	0.07	0.41	0.30
—per dollar of gross revenue[a]	n/a	$0.67	$0.54	$0.43	$0.31
—per ton of CO_2 reduced[b]	$16.17	$19.97	$16.28	$42.12	$31.30
Climate benefits[c]					
—Average, 3% discount rate	$2,582.41	$2,517.76	$2,517.40	$5,678.74	$5,646.42
Co-benefit estimates					
—Energy security premium[d]	$0.23	$0.02	$0.06	$74.40	$74.61
—Local pollution benefits[e]	$3,282.86	$3,561.85	$3,511.27	$8,454.59	$8,091.67
—NO_x	$213.82	$228.03	$225.72	$700.85	$684.20
—$PM_{2.5}$	$702.69	$805.30	$754.92	$3,336.61	$3,009.45
—SO_2	$2,366.34	$2,528.51	$2,530.62	$4,417.13	$4,398.02
Total benefits	$5,865.50	$6,079.62	$6,028.73	$14,207.74	$13,812.70

[a] Present value of gross revenue, in 2013 dollars.

[b] Welfare cost divided by present value of cumulative tons reduced, using reference case interest rate.

[c] In billions of 2013 dollars. Present value of emissions reductions times social cost of CO_2 in 2013 dollars.

[d] In billions of 2013 dollars. Present value of reduced oil imports, valued at $5.33 per barrel.

[e] In billions of 2013 dollars. Present value of cumulative reductions, valued at EPA estimates of benefits per ton reduced by source.

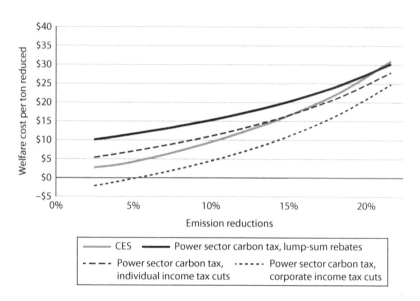

FIGURE 6.6

Welfare Costs of CES versus Power-Sector Carbon Tax

Figure 6.6 displays the welfare costs of the CES and power-sector-only carbon tax as a function of stringency.[8] For all policies, costs per ton rise with stringency. In some cases, the relative costs change with stringency, while in others, they do not.

When the power-sector-only carbon tax policy involves recycling through cuts in individual income taxes (on wages, dividend, interest, and capital gains), the relative cost-effectiveness depends on stringency. The CES is less costly when the policy is relatively lax. At very low levels of stringency, the CES's advantage in terms of producing a smaller tax-interaction effect more than offsets its disadvantage in terms of its inability to elicit efficient ratios of coal-fired, other-fossil-, and nonfossil-generated electricity. That disadvantage grows with stringency and becomes more important relative to the tax-interaction effect.[9] As the policy becomes more stringent and high levels of abatement become necessary, the CES's disadvantage in terms of fuel substitutions becomes sufficiently important that the CES becomes more costly than the carbon tax policy.

When the power-sector-only carbon tax policy involves lump-sum recycling, costs per ton are always lower under the CES, reflecting the fact that with lump-sum recycling the carbon tax policy does not exploit the revenue-recycling effect. When the power-sector-only carbon tax takes full advantage of revenue recycling by returning revenues through cuts in corporate income taxes,[10] the carbon tax's disadvantage in terms of the tax-interaction effect is outweighed by the strong revenue-recycling benefit as well as the carbon tax's ability to elicit more efficient ratios of coal-fired, other-fossil-, and nonfossil-generated electricity.

Benefits: Climate and Nonclimate Benefits

Table 6.2 also displays the climate and nonclimate benefits of the central-case medium stringency CES policy (with partial credit for natural gas) and the equivalent power-sector-only carbon tax.[11] The climate benefits of both policies greatly exceed the welfare costs.[12] Neither policy produces significant energy security benefits related to reduced oil imports because the electricity sector uses very little oil (heating oil, more specifically) to produce electricity. The benefits from reduced $PM_{2.5}$ pollution and $PM_{2.5}$ precursors, NO_x and SO_2, also greatly exceed the welfare costs of each policy. In fact, under each policy the health-related benefits from reduced $PM_{2.5}$ pollution alone exceed the welfare costs. Both the CES policy and the equivalent power-sector-only carbon tax with individual income tax cuts generate total benefits that are 10 times greater than the welfare costs.

Results Under Alternative Specifications

The absolute and relative costs of the CES depend on the size of the tax-interaction effect. As indicated in chapter 2, the tax-interaction effect is larger, the higher the preexisting taxes are on labor and capital. Here we alter the magnitude of the tax-interaction effect by performing counterfactual simulations in which the preexisting marginal rates for individual income taxes (on wages, interest, dividends, and capital gains) and corporate income taxes are 25 percent lower or 25 percent higher than in the central case.

Results are displayed in figure 6.7. As was the case for emissions pricing, the welfare costs of the CES policy increase with the level of preexisting taxes (figure 6.7a), a symptom of a larger tax-interaction effect. However, the ratio of the CES cost to the cost under emissions pricing

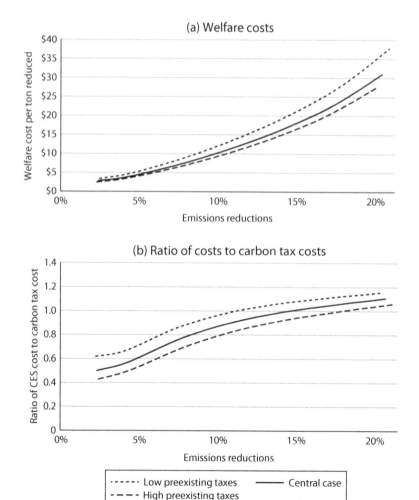

FIGURE 6.7

Preexisting Taxes and Welfare Costs of CES Policies

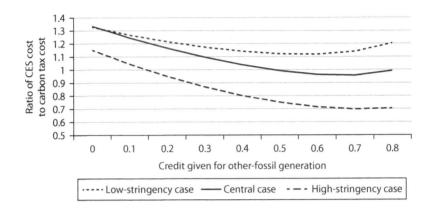

FIGURE 6.8

Relative Welfare Costs of CES Policies under Alternative Natural Gas Credits

decreases with the magnitude of preexisting taxes (figure 6.7b), a consequence of the fact that emissions pricing has a larger tax-interaction effect than intensity standards.[13]

Figure 6.8 reveals the significance of the size of the credit to natural-gas-generated electricity to the relative costs of the CES and the carbon tax policies, when the latter involves individual income tax recycling.[14] In the absence of any credit, the relative share of natural-gas-generated electricity by utilities will be insufficient from an efficiency point of view: relatively low-cost reductions of greenhouse gas emissions could be accomplished through greater use of natural gas. However, if natural-gas-generated electricity receives credits in excess of 0.6, then there is too much natural-gas-generated electricity and it would be efficient to reduce its use and rely more on nonfossil generation to reduce emissions. This ideal credit of 0.6 is consistent with the relative emissions factors of natural-gas-generated electricity and coal-fired electricity: 1 megawatt hour produced from natural gas produces about 60 percent less carbon dioxide than 1 megawatt hour produced from coal.

Table 6.3 displays the welfare costs under the CES as well as the costs of the CES relative to those under a power-sector carbon tax, for policies achieving cumulative emissions reductions of 10 percent, 15 percent, and 20 percent. Results are shown under the central case and alternative

TABLE 6.3 CES Welfare Costs and Relative Welfare Costs per Ton of Emissions Reduction Under Alternative Parameter Assumptions

| | POLICY STRINGENCY | | | | | |
| | 10% | | 15% | | 20% | |
	CES	RATIO OF CES TO EQUIVALENT CARBON TAX[a]	CES	RATIO OF CES TO EQUIVALENT CARBON TAX[a]	CES	RATIO OF CES TO EQUIVALENT CARBON TAX[a]
Central case	$9.65	0.86	$16.92	1.00	$27.37	1.08
Elasticity of labor supply (φ) —central value = 0.3						
$\varphi = 0.1$	$8.10	0.81	$14.35	0.97	$23.37	1.06
$\varphi = 0.5$	$11.72	0.90	$20.35	1.03	$32.67	1.11
Intertemporal elasticity of substitution ($1/\sigma$) —central value = 0.5						
$1/\sigma = 0.25$	$9.90	0.92	$16.84	1.02	$26.81	1.08
$1/\sigma = 0.75$	$9.56	0.83	$16.96	0.99	$27.61	1.08
Adjustment costs (ξ) —central value = 7						
$\xi = 2$	$8.24	0.78	$15.23	0.96	$25.33	1.08
$\xi = 14$	$11.56	0.98	$19.19	1.07	$30.18	1.13
Generator elasticity of substitution (σ_{gen}) —central value = 3						
$\sigma_{gen} = 2$	$13.17	0.99	$23.12	1.15	$37.01	1.24
$\sigma_{gen} = 4$	$8.34	0.83	$14.53	0.95	$23.89	1.03

[a] Equivalent carbon tax defined as electricity-only carbon tax with personal income tax cuts that achieve same aggregate emissions reductions as CES policy.

parameter values considered in chapter 5, and recycling is via individual income tax cuts. As discussed in chapter 5, the tax-interaction effect is increasing in the elasticity of labor supply and the intertemporal elasticity of substitution. Although the tax-interaction effect is relatively small for the CES policy, the welfare costs of a given level of cumulative economy-wide emissions reductions is increasing in these parameters. Hence one might expect the CES policy to perform better than the equivalent carbon tax with higher values for the elasticity of labor supply and the intertemporal elasticity of substitution, but that is not the case. The revenue-recycling effect is also increasing in these elasticities. As a result, under individual income tax recycling, the relative costs of the CES are lowest at the lower elasticity values.

Adjustment costs mainly influence the relative costs of the CES policy by affecting utilities' ability to substitute across generators. If adjustment costs are low, then changes in investment will quickly lead to changes in the generation mix through changes in capital stocks of each generator. Because CES relies primarily on substitution between generators to reduce emissions, lower (higher) adjustment costs lead to lower (higher) welfare costs of the CES policy. Lower adjustment costs will also increase the amount of optimal substitution relative to output reduction under a carbon tax. But because the CES policy attaches greater weight to substitution, greater substitutability across time through lower adjustment costs favors the CES policy.

Finally, the relative costs also depend on the intratemporal elasticity of substitution between generators, which directly affects the ease with which utilities can substitute between coal-fired, natural-gas-fired, and nonfossil-generated electricity. As with adjustment costs, both the level of the CES's welfare costs and its costs relative to those under a power-sector carbon tax are decreasing in this elasticity. A higher value for this elasticity implies a lower cost of reducing emissions by substituting from dirtier to cleaner generators. This lowers the cost under both policies. As discussed, the CES relies more heavily on this emissions reduction channel than does emissions pricing, which (because of higher electricity prices) makes greater use of the channel of reduced overall demand for electricity. As with intertemporal substitution, greater intratemporal substitution enhances the relative attractiveness of the CES policy. Quantitatively, the relative welfare costs depend more on the intratemporal generator elasticity of substitution than on the intertemporal elasticity through changes in adjustment costs.

Summary

The oft-claimed superiority of emissions pricing instruments over intensity standards on cost-effectiveness grounds does not always hold: tax interactions can reverse the rankings. Prior economic research shows that intensity standards such as the CES generally fail to produce the most efficient mix of inputs and also fail to achieve the most efficient extent of demand-side conservation. We find, however, circumstances in which the CES's advantages in terms of a lower tax-interaction effect can fully compensate for these limitations.

The CES's tax-interaction advantage is especially important when the emission reductions called for are modest. As policies become more stringent, its disadvantages relative to emissions pricing gain relative importance. This does not contradict the finding in chapter 5 that the costs of emissions pricing can become very small—even negative—when the price of carbon is very low and emissions abatement is modest. What the present analysis indicates is that the CES can involve even lower costs than the low costs from carbon pricing. The potential of the CES to involve lower costs than those under carbon pricing is expanded to the extent that the revenues from carbon-pricing policies are not recycled in an efficient manner.

INCREASING THE FEDERAL GASOLINE TAX

The federal tax on gasoline is currently 18.4 cents per gallon. All states also impose taxes on gasoline, from a low of 9 cents per gallon in Alaska to a high of 51.4 cents per gallon in Pennsylvania; the weighted average of these state-level taxes is 26.5 cents per gallon.[15] On average, gasoline is taxed at a rate of about 44.9 cents per gallon in the U.S. Here we consider increases in the federal component of 15, 30, and 60 cents per gallon.

Although gasoline combustion contributes importantly to U.S. CO_2 emissions, the coverage of the gasoline tax policy would be considerably narrower than that of the emissions pricing policies considered earlier.

While emissions pricing policies have the potential to cover most U.S. CO_2 emissions, the gasoline tax policy would likely cover less than a quarter. In 2013, about 20 percent of U.S. emissions of CO_2 emanated from combustion of gasoline across sectors. This means that very large increases in the gasoline tax would be required to achieve CO_2 emissions reductions comparable to the central case carbon tax or cap-and-trade policies we have considered. The simulation results from the gasoline tax increases described later bear this out.

However, other potential impacts of a gasoline tax increase could lend support to this policy. One consideration is national security. Increases in the gasoline tax could have a significant impact on the nation's demands for petroleum and for imported oil in particular. Several analysts claim that reduced reliance on oil imports adds to national security.[16] A gasoline tax also has attractions in terms of its implications for pollution and road congestion. To the extent it causes people to drive less, it yields reduced automobile-generated emissions of local pollutants (including carbon monoxide and nitrogen oxides), less road congestion, and fewer accidents. Parry and Small (2005) find that the optimal gasoline tax in year-2000 dollars is about $1.00. Approximately 69 percent of that optimal tax comes from the value of reduced local pollutants, road congestion, and accidents; only 5 percent derives from the value of reduced CO_2 emissions. Therefore, the nonclimate benefits from increased gasoline taxes could dwarf the gains they bring about from reduced carbon emissions.[17]

However, a carbon tax also would yield nonclimate benefits (as shown in chapter 5). A priori, it is not evident how the overall (climate and non-climate) gains from a gasoline tax compare with those of a carbon tax. In this chapter we apply the E3 model to make this comparison.

We start by focusing on the potential of increased gasoline taxes to reduce CO_2 emissions and on the costs of reducing CO_2 emissions via this policy. We compare the impacts of the gasoline tax increase with those of the climate policies already examined in terms of overall economic cost and impacts on households and industries. Toward the end of the chapter, we also take a brief look at the impacts of gasoline tax increases on local pollution emissions and oil imports to give an approximate sense of the possible "co-benefits." Using the information on this policy's economic costs and co-benefits, we then compare the overall net benefits of the gasoline tax increase with those from a carbon tax that yields the same emissions reductions.

In evaluating an increase in gasoline taxes, it is important to recognize that the magnitude of its co-benefits depends on the extent to which policy makers have implemented other policies that could target local pollution, congestion, and imports. If these other (perhaps better targeted) policies were in place, the gasoline tax's co-benefits would be more limited. Hence it is appropriate to regard a gasoline tax increase as a "second-best" approach to addressing local pollution, congestion, and oil imports.

Modeling the Gasoline Tax

The gasoline tax is implemented as a per-unit tax on the consumption of the good in the category "motor vehicle fuels." While we call this policy an increase in the gasoline tax, it is in fact an increase in the taxes on both gasoline and diesel. Recall from chapter 3 that for any consumer good, the after-tax price $\tilde{p}$ is a function of the ad valorem tax τ_{ca}, the excise tax τ_{ce}, the income tax deduction s^d, the average individual income tax rate that applies to deductions τ_s, and the level of tax credits s^c:

$$\tilde{p}_j = (1 - \tau_s s_j^d)[(1 + \tau_{ca,j})\hat{p}_j + \tau_{ce,j}] - s_j^c$$

Motor vehicle fuel purchases are subject to neither ad valorem taxes nor tax credits and deductions; therefore, the price of motor vehicle fuels (good 10) simplifies to $\tilde{p}_{10} + \tau_{ce,10}$. The gasoline tax is modeled as an increase in the excise tax $\tau_{ce,10}$. The tax rate is constant in real terms; hence, the nominal rate rises at the rate of inflation (Consumer Price Index) over time. The tax applies to only the household use of motor vehicle fuels. It does not apply to industrial uses of gasoline or diesel, including uses for air, water, or truck transportation.

Design Dimensions

We consider three permanent one-time increases in the federal tax on gasoline: 15, 30, and 60 cents per gallon in 2013 dollars. The policy is introduced in 2017, as was the case with the other climate policies.

Like the other policies, this policy is revenue-neutral. This neutrality is achieved either through lump-sum rebates to households or via cuts in the marginal rates of individual taxes on labor and capital income.

Central Case Results

Emissions Reductions

Figure 6.9 displays the emissions impacts of the three gasoline tax increases. In 2030, the emissions reductions are 0.4, 0.7, and 1.4 percent of baseline emissions, under the 15-, 30-, and 60-cent gas tax increases, respectively. These reductions are roughly proportional to the size of the tax increase. Compared with the emissions reductions under the climate policies previously examined, the magnitude of these reductions is small. In the case of the largest (60-cent) increase in the gas tax, emissions are reduced by about 75 million metric tons annually. In contrast, our central case carbon tax yielded reductions about 10–30 times larger: in 2019 (when the carbon tax is $20 per ton), the tax produces reductions of about 840 million metric tons, and in 2050 (when it has reached $60 per ton) it yields reductions of over 2,000 million metric tons.

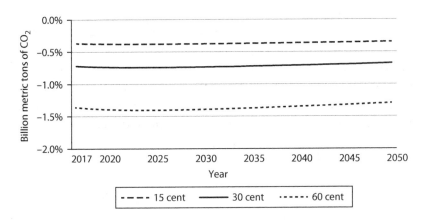

FIGURE 6.9

Emissions Reductions from Increased Gasoline Taxes

The gasoline tax's smaller impact mainly reflects its much narrower base. In the reference case, emissions from household use of motor vehicle fuels account for 9-10 percent of total U.S. carbon emissions in the years 2017–2050. At the same time, a 60-cent gasoline tax increase corresponds to an extra tax of about $70 per ton of the carbon in the gasoline—not much different from the carbon tax rates considered earlier.[18] Hence the differences in emissions reductions are mainly due to differences in the tax base rather than differences in the rate at which carbon is taxed.

Prices, Consumption, and Revenues

As mentioned in chapter 3, the world price of oil is exogenous in the E3 model. In our central case, this price rises by $1.24 (in 2013 dollars) each year. The world price determines the price of crude oil to domestic users (until oil imports go to zero). The rising crude oil price is then reflected in gasoline prices that increase over time. Figure 6.10a displays the time profiles of the prices of crude oil and of gasoline, relative to the price in the benchmark year 2013. The gasoline prices are shown for

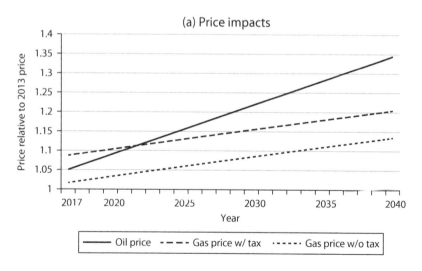

FIGURE 6.10

Price, Consumption, and Revenue Impacts of a Gasoline Tax Increase

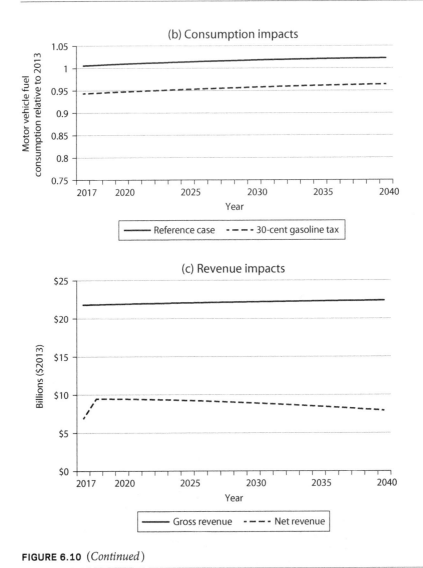

FIGURE 6.10 (*Continued*)

both the reference case and the case of the 30-cent increase in gasoline taxes. The constant increase in the tax implies a parallel upward shift in the time line for the gasoline price.

The rising prices promote reductions in the demand for gasoline, other things equal. As indicated in figure 6.10b, gasoline consumption grows at a rate of about 0.1 percent, despite the continued exogenous growth of the overall economy of 1 percent annually. The impacts on gasoline consumption under the 15- and 60-cent gasoline tax increases are similar and roughly proportional to the magnitude of the gasoline tax increase.[19]

Correspondingly, the gross and net revenue impacts of the gasoline tax increase are fairly constant over time, as shown in figure 6.10c for the 30-cent gasoline tax increase. The impacts for the 15- and 60-cent increases are roughly 50 percent smaller and larger, respectively. As was the case for the other climate policies, net revenues are significantly below gross revenues, reflecting the adverse impact on the tax base. For the 30-cent gasoline tax increase, over time net revenues are consistently about 40 percent of gross revenues, with the exception of the first period.[20]

GDP, GDP Components, and Welfare

Table 6.4 presents the impacts on GDP and its components for selected years. The GDP costs are smaller than those of the other climate policies. This is as expected, given this policy's much smaller coverage.

Under recycling via lump-sum rebates, the impacts on GDP, consumption, and investment are negative and approximately constant over time. Consumption falls by more than investment, in keeping with the fact that the gasoline tax—a tax on a consumer good—has raised the price of consumption relative to investment.

TABLE 6.4 **Impacts on GDP and GDP Components of a 30-Cent Increase in the Gasoline Tax**
Percentage Changes from Reference Case

	LUMP-SUM RECYCLING			INDIVIDUAL INCOME TAX RECYCLING		
	2020	2030	2040	2020	2030	2040
GDP	−0.03	−0.03	−0.03	−0.02	0.00	0.01
Consumption	−0.05	−0.04	−0.03	−0.04	−0.02	0.00
Investment	−0.01	−0.02	−0.02	0.05	0.04	0.04

In contrast with the lump-sum rebate case, under recycling via cuts in the individual income tax rate, the impacts on GDP and consumption change over time. The cut in the individual income tax reduces the tax rate on components of capital income—dividends, interest, and capital gains. The combination of the gasoline tax and the individual income tax rate reductions increases the attractiveness of investment. Thus, although the GDP impact is negative in the short term, over time the effect on GDP eventually becomes positive, reflecting the increase in investment.

Welfare impacts are displayed in table 6.5. For comparability, the gasoline tax policy's impacts are compared with those of a carbon tax that has been scaled to generate the same cumulative CO_2 emissions reductions.[21] The table shows that, for any given form of recycling, the gasoline tax involves significantly higher welfare costs than those of the equivalent carbon tax. Because of the gasoline tax's narrower base, relatively high rates per ton of carbon are needed to achieve the same reductions as those under the carbon tax.[22] The higher rates cause larger economic distortions and thus imply significantly higher economic costs (though the overall costs of both policies are low compared with those of the previously considered policies).

The table also shows a pattern that has emerged throughout the policy analyses of this book: recycling in the form of cuts in individual income tax rates involves significantly lower costs than in the case of recycling via lump-sum rebates.

Figure 6.11 indicates, for a range of policy stringencies, the welfare costs per ton under the gasoline tax and similarly scaled carbon tax. As indicated in the figure, the costs per avoided ton are substantially higher under the gasoline tax policy than under the carbon tax. Costs rise as the stringency of the gasoline tax increase, but the percentage increase is slight because of the preexisting gasoline tax and prior taxes on factors of production.

Climate and Nonclimate Benefits

As in chapter 5's analysis of the carbon tax, the climate benefits are based on the central case SCC estimates ($43 per ton in 2017, rising to $78 per ton in 2050, in 2013 dollars). Table 6.5 includes estimated climate benefits under the central case gasoline tax increase, the comparably scaled carbon tax, and the central case carbon tax from chapter 5. By construction, the

TABLE 6.5 Costs and Benefits of a Federal Gasoline Tax Increase and a Carbon Tax

	30-CENT GASOLINE TAX INCREASE		CARBON TAX—MATCHING PV OF EMISSION REDUCTIONS		CENTRAL CASE CARBON TAX	
	LUMP-SUM REBATES	CUTS IN INDIVIDUAL INCOME TAX	LUMP-SUM REBATES	CUTS IN INDIVIDUAL INCOME TAX	LUMP-SUM REBATES	CUTS IN INDIVIDUAL INCOME TAX
Welfare costs	$285.75	$165.38	$28.66	$13.03	$3,356.47	$2,479.64
—as percent of wealth	0.03	0.02	0.00	0.00	0.41	0.30
—per dollar of gross revenue[a]	$0.30	$0.17	$0.27	$0.12	$0.43	$0.31
—per ton of CO_2 reduced[b]	$160.06	$98.66	$16.05	$7.30	$42.12	$31.30
Climate benefits[c]						
—Average, 3% discount rate	$121.19	$113.26	$124.50	$124.46	$5,678.74	$5,646.42
Co-Benefit estimates						
—Energy security premium[d]	$10.61	$10.65	$1.12	$1.14	$74.40	$74.61
—Local pollution benefits[e]	194.81	140.44	192.43	186.81	$8,454.59	$8,091.66
—NOx	$60.67	$58.01	$14.28	$14.04	$700.85	$684.20
—PM$_{2.5}$	$122.29	$75.71	$54.07	$48.36	$3,336.61	$3,009.45
—SO$_2$	$11.85	$6.72	$124.08	$124.41	$4,417.13	$4,398.02
Total benefits	$326.60	$264.36	$318.05	$312.41	$14,207.74	$13,812.70

a Present value of gross revenue in 2013 dollars.
b Welfare cost divided by present value of cumulative tons reduced, using reference case interest rate.
c In billions of 2013 dollars. Present value of emissions reductions times social cost of CO_2 in 2013 dollars.
d In billions of 2013 dollars. Present value of reduced oil imports, valued at $5.33 per barrel.
e In billions of 2013 dollars. Present value of cumulative reductions, valued at EPA estimates of benefits per ton reduced by source.

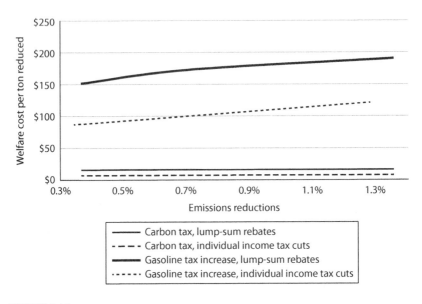

FIGURE 6.11

Welfare Costs of the Gasoline Tax Increase

climate benefits are virtually the same for the comparably scaled policies.[23] Under the gasoline tax, these benefits are less than the welfare costs. In contrast, for the comparably scaled carbon tax, although the climate benefits are similar to those of the gasoline tax increase, these benefits exceed the welfare costs because the latter are considerably lower.

By discouraging gasoline consumption, the gasoline tax policy reduces demands for petroleum and for imported oil in particular. Figure 6.12 displays the time profile of oil imports in the reference case and under the 30-cent gasoline tax with lump-sum recycling. In both the reference case and the policy case, imports rise initially in keeping with the declining output of the oil sector, but imports begin to fall as the backstop technology grows over time.

The analysis in chapter 5 of the carbon tax included estimates of potential energy security benefits associated with reduced imports of oil. Those benefits were based on an energy security premium of $5.33 per barrel (in 2013 dollars). Applying the same value for the energy security premium here, we find that the 30-cent gasoline tax increase would generate an

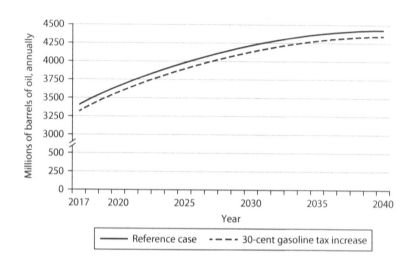

FIGURE 6.12

Oil Imports and the Gasoline Tax Increase

energy security value of $10.6 billion (in present value) from reduced oil imports.[24] The equivalent carbon tax would yield an energy security value of only $1.1 billion. Thus the gasoline tax increase has an advantage over a comparably scaled carbon tax in terms of energy security benefits. This advantage does not outweigh its cost disadvantage, however.

The combustion of gasoline and diesel fuel in automobiles emits a number of air pollutants. Table 6.6 shows, for each criteria air pollutant, the differences in emission reductions between the gasoline tax increase and a comparably scaled carbon tax. With the exception of SO_2, the gasoline tax increase produces larger reductions in the local pollutants.

As in chapter 5, we consider the monetized value of the avoided health impacts stemming from reduced emissions of some of the local pollutants. We noted previously that estimates are not available for many pollutants, including CO and VOCs. Hence, as in chapter 5, we focus on the available monetized value of health benefits related to reduced $PM_{2.5}$ pollution.[25]

Table 6.5 displays these benefits, in billions of 2013 dollars. The overall values of these co-benefits from a 30-cent increase in the federal gasoline tax are $194 billion and $140 billion under lump-sum rebates and individual income tax recycling, respectively. Most of the value stems from

TABLE 6.6 Criteria Pollutant Emissions: Gasoline Tax vs. Carbon Tax
Percent Changes from Reference Case

POLLUTANT	2020			2030			2050		
	GASOLINE TAX (%)	EQUIVALENT CARBON TAX (%)	DIFFERENCE (IN PERCENTAGE POINTS)	GASOLINE TAX (%)	EQUIVALENT CARBON TAX (%)	DIFFERENCE (IN PERCENTAGE POINTS)	GASOLINE TAX (%)	EQUIVALENT CARBON TAX (%)	DIFFERENCE (IN PERCENTAGE POINTS)
CO	−2.04	−0.05	−1.99	−1.80	−0.06	−1.74	−1.34	−0.05	−1.29
NO$_x$	−1.16	−0.23	−0.93	−1.06	−0.29	−0.78	−0.83	−0.26	−0.57
PM$_{10}$	−0.11	−0.04	−0.07	−0.11	−0.05	−0.06	−0.08	−0.04	−0.04
PM$_{2.5}$	−0.14	−0.07	−0.07	−0.14	−0.08	−0.05	−0.10	−0.07	−0.03
SO$_2$	−0.05	−0.93	0.89	−0.08	−1.30	1.21	−0.09	−1.38	1.29
VOC	−0.34	−0.02	−0.31	−0.31	−0.03	−0.28	−0.22	−0.02	−0.20
NH$_3$	−0.19	−0.02	−0.17	−0.18	−0.02	−0.16	−0.13	−0.02	−0.11

reductions in $PM_{2.5}$ emissions, which are much larger under the gasoline tax increase than from the equivalently scaled carbon tax. In contrast, the gasoline tax increase leads to much smaller reductions in SO_2 than does the equivalent carbon tax. Taken together, the health benefits from the three pollutants we have considered are very similar to those under the carbon tax when recycling is via lump-sum rebates. The gasoline tax's local benefits are smaller, however, when recycling is through cuts in individual income tax rates. The economic boost from reduced taxes increases activity associated with local air pollution in other sectors. The same holds true for the economywide carbon tax, but the broad coverage of the carbon tax prevents significant increases in criteria air pollutants due to tax cuts.

Summary

A 30-cent increase in the federal gasoline tax would represent a 67 percent increase in the tax relative to the average total (federal plus state average) gasoline tax under the status quo. The implicit charge per ton of CO_2 under this policy is roughly similar to the average price of CO_2 under the central case carbon tax from chapter 5. But the impact on emissions is much smaller: the gasoline tax increase yields only one 50th of the emissions reductions under the central case carbon tax. This reflects the considerably narrower base of the gasoline tax and the inelasticity of gasoline demand relative to the average elasticities across various sectors of CO_2 emissions with respect to actual or implied CO_2 prices.

We have also compared the gasoline tax increase with a (lower) carbon tax that yields the same cumulative CO_2 emissions reductions. Because of its narrower base, the gasoline tax policy's costs per ton are approximately 10 times greater than the equivalently scaled carbon tax. Thus, the carbon tax emerges as the much more cost-effective instrument for achieving CO_2 reductions.

However, the two policies differ in terms of co-benefits. Accordingly, we have also compared the policies while taking the co-benefits into account. We have concentrated on the co-benefits in the form of avoided health costs from air pollution and enhanced national security from reduced imports of oil. In value terms, the health benefits from avoided air pollution are much larger than the national security benefits. Compared with the equivalently scaled carbon tax, the gasoline tax yields higher benefits from reductions in $PM_{2.5}$, while offering smaller benefits from reductions in SO_2. The health benefits from the reductions in emissions of $PM_{2.5}$, SO_2, and NO_x combined are roughly similar under the two policies when the policies recycle their revenues via lump-sum rebates.

The net benefits from the gasoline tax increase are positive under both forms of recycling considered, but the magnitudes depend significantly on the form of recycling. Under revenue recycling via lump-sum rebates, the net benefit is $41 billion—the difference between total benefits (climate benefits plus co-benefits) of $327 billion and costs of $286 billion.[26] When revenue recycling takes the form of cuts in individual income taxes, the net benefit is $100 billion—total benefits are $265 billion and costs are $165 billion. The higher net benefits in this latter case reflect the significantly lower costs under recycling through individual income tax cuts.

The benefit-cost ratio of an equivalently scaled carbon tax is considerably higher than that under the gasoline tax increase. This is the case because the carbon tax involves significantly lower costs, a reflection of its much broader base.

CONCLUSIONS

Our explorations of the two alternative policies offer contrasting results. In the case of the CES, attention to tax interactions strengthens the case for the CES relative to the emissions pricing policies considered in chapter 5. Indeed, if the emissions reduction target is not very stringent, the target can be reached at lower cost under the CES than under emissions pricing. Here the CES's advantage in terms of producing a smaller tax-interaction effect outweighs its disadvantages along other dimensions that affect cost. Under low levels of policy stringency, the CES's cost-advantage applies even when the equivalent emissions pricing policy incorporates a relatively efficient form of revenue recycling—cuts in individual income tax rates. When revenues from carbon pricing policies are recycled relatively inefficiently, the CES has a cost-advantage at low to moderate levels of stringency. As with emissions pricing policies, the benefits from the CES policy significantly outweigh the costs: the benefits of the Bingaman-type CES policy outweigh its costs by a factor of 10.

In contrast, attention to tax interactions does not improve the case for increased gasoline taxes as an instrument for reducing CO_2 emissions. Such interactions have roughly similar impacts on the costs of gasoline tax increases and equivalently scaled carbon taxes. More critical to policy costs is the gasoline tax's narrow base, which underlies the gasoline tax's considerably higher costs per ton of reduced CO_2 compared with a carbon tax. Increasing the gasoline tax yields considerable co-benefits in the form of avoided health damages from reduced air pollution. However, a carbon tax scaled to achieve comparable reductions in CO_2 emissions gives rise to co-benefits of roughly similar magnitude.

7

DISTRIBUTION OF POLICY IMPACTS
ACROSS INDUSTRIES AND HOUSEHOLDS

Previous chapters focused primarily on climate policies' impacts on the economy as a whole. However, the impacts usually differ across industries and households. The distribution of economic losses (or gains) across industries or households depends on the design of the policy involved. Because decisions about policy design affect the distribution of impacts, these decisions have important implications in terms of fairness. They also have important implications for political feasibility: clearly, to the extent that costs can be reduced for groups wielding sufficient political power to block a policy effort, the prospects for political success are enhanced.

This chapter examines the distribution of impacts of climate-change policies across industries and household income groups. We focus primarily on the emissions pricing policies from chapter 5—a carbon tax and a cap-and-trade system—although we give some attention to the other policies we addressed earlier. We show how climate policies can be formulated to soften the impacts on sectors or household groups that might otherwise absorb an especially large share of the policy costs.

In most cases, achieving a more even distribution of economic impacts, whether across industries or across household income groups, comes at a cost: there are trade-offs between cost-effectiveness and distributional equity. There is a trade-off between cost-effectiveness and political feasibility as well to the extent that lowering the burdens on influential stakeholders (and thereby building their political support) raises economic costs. This chapter applies the E3 model to assess these trade-offs.

We begin by focusing on the differing impacts on industry profits. We show how policies can be designed to avoid losses of profit in carbon-intensive industries and how those policy designs affect the overall economic costs of the policy.

We then consider the impact of climate policies across household income groups through effects on labor and capital incomes and effects on the prices of the goods and services purchased. We explore a range of policy designs in order to assess the economic costs of preventing losses to certain groups or of achieving a more even distribution of impacts across income categories.

IMPACTS ACROSS INDUSTRIES

Chapter 5 showed that certain emissions pricing policies lead to significant reductions in the profits of some industries. Table 5.4 revealed this for our central case carbon tax. The profit impacts were identical under an equivalently scaled cap-and-trade program, that is, one in which the caps are set in each year to match the levels of emissions that result from the carbon tax. Under these policies, the profit losses were largest in percentage terms for energy industries: coal-fired electricity generation and coal mining faced the largest losses, as coal has the highest carbon content and there exist substantial possibilities for substitutions in the electric power sector. Industries that produce, consume, or transport oil and natural gas also suffered large profit losses.[1]

Distribution of Profit Impacts Under Cap and Trade

As discussed in chapter 5, the free allocation of emissions allowances under cap and trade can completely offset the negative profit impacts on carbon-intensive industries, compared with the case where allowances are auctioned out. Indeed, if enough allowances are given out free, cap and trade can produce significant economic rents and thereby cause profits to rise in these industries.

In the examination of the profit impacts of free allocation in chapter 5, we focused on the limiting case of 100 percent free allocation. And we assumed a particular way of allocating the free allowances: they were allocated either based on historical emissions or in proportion to the E3 model's predicted profit losses under the central cap-and-trade program with lump-sum revenue recycling. However, there are many other design options in terms of the fraction of total allowances given out free and the shares of the free allowances going to particular industries. Here we will explore how both of these dimensions affect the distribution of profit impacts under cap and trade. We consider, in particular, how a cap-and-trade program can be designed to eliminate adverse profit impacts on industries that are most vulnerable to large impacts on profits.

Providing free emissions allowances is not the only way to help certain industries avoid losses of profit. Another option is targeted (industry-specific) tax relief, a form of compensation that could be financed through proceeds from the sale of emissions allowances. We will focus on targeted tax relief later as part of our exploration of distributional impacts of a carbon tax. Many of the insights from that discussion also apply to cap and trade.

Allowance Allocation, Rent and Profit: A Graphical Illustration

To illustrate key connections between the method of allowance allocation and industry profits, we offer a figure here that conveys some of the basic theory.[2] The example illustrated by figure 7.1 applies to firms that must hold and submit allowances. It is assumed that the firms are within a competitive industry.

In the absence of regulation, the equilibrium price and level of output at the industry level are shown as p_0 and X_0, respectively. These are determined by the intersection (at point b) of the original supply and demand curves S_0 and D.

Introducing a cap-and-trade system covering emissions from this industry gives firms incentives to reduce emissions. Regardless of whether allowances initially are auctioned out or given out free, each additional unit of emissions comes at a cost to the emitter, either obliging the firm to purchase an additional allowance in the market or reducing the number of surplus allowances that the firm can sell. Firms generally can accomplish the emissions reductions by changing the input mix (e.g., switching to less carbon-intensive fuels) or reducing levels of output.[3] In figure 7.1 it is assumed that fuel switching raises per-unit production costs by c for firms in this industry. The figure also assumes that the cost per unit of remaining emissions—the allowance price multiplied by emissions per unit of output—is r. The policy thus yields the new industry supply curve S_1.

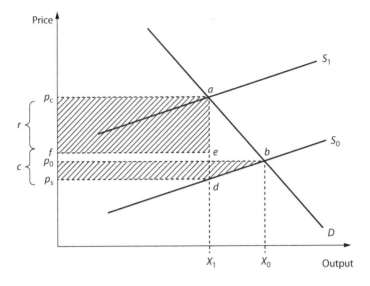

FIGURE 7.1

Allowance Allocation, Rents, and Profits

In the new equilibrium, output is X_1 and the consumer price p_C exceeds the original marginal supply cost by $c + r$.

If the allowances are introduced through a competitive auction, the policy will generate no rents: they will be bid away through competitive bidding for the limited number of allowances. In this case, the loss of producer surplus is the shaded trapezoid p_0bdp_S, while the loss of consumer surplus is p_Cabp_0. The shaded rectangular area p_Caef represents the revenue that the government receives from the allowance auction. These revenues can then be used to provide rebates to households or to finance cuts in preexisting taxes on capital and labor income. Alternatively, the revenues p_Caef could be used to pay for additional public spending, in which case the general public would benefit from the goods or services provided.

If instead the allowances are introduced through free allocation, the distribution of impacts between producers and taxpayers is very different. In this case the shaded rectangle p_Caef represents rents to producers rather than revenues to the government. In the figure, these rents are about three times as large as the *gross* loss of producer surplus (the loss before considering the rents) represented by the shaded area p_0bdp_s. Thus, in this figure, free allocation enables the firm to enjoy a higher profit than in the absence of regulation.[4]

Another option is combining free allocation and auctioning. For the industry depicted in the diagram, industry profits would be preserved if about one-third of the allowances were given out free, thus enabling the firm to enjoy about a third of the potential rents. The rest of the allowances could be auctioned out. Relative to free allocation, auctioning has advantages in terms of cost-effectiveness to the extent that the revenues from the auction are used to finance cuts in marginal tax rates.

As suggested by figure 7.1, the amount of free allocation necessary to preserve a given firm's profits depends on the extent to which the firm can shift the burden of regulation to consumers. That depends on the relative price elasticity of supply and demand. A higher relative elasticity of supply (a supply curve flatter than the demand curve) implies a larger pass-through of compliance costs into producer prices and a smaller gross loss of producer surplus. It also implies larger potential rents relative to the gross loss of producer surplus, which in turn means that less free allocation (or rents to be retained by firms) is needed to maintain profits.

Allowance Allocation and Profits: Numerical Results

The analytical framework just described enables us to understand what underlies the differing impacts of cap and trade under alternative specifications for allowance allocation. Here we contrast results from 100 percent auctioning and 100 percent free allocation—cases considered in chapter 5— with results from a new hybrid case in which some allowances are freely allocated to key industries and the remaining allowances are auctioned. In this hybrid case, the free allowances are allocated to the ten industries that would experience the largest percentage reductions in profit under the 100 percent auctioning policy.[5] Just enough allowances are freely allocated to each industry to ensure that the present value of profits (over an infinite time horizon) is unchanged from the reference case. We call the allocation in the hybrid case *profit-preserving free allocation.*

The potential burden of the cap-and-trade policies is not confined to firms in the industries selected as points of regulation. Firms in these industries generally will shift some of the burden forward to their customers and backward to their suppliers. Firms in some industries outside the points of regulation are among the most vulnerable. Thus the set of industries receiving free allowances in the profit-preserving free allocation case includes industries that are not points of regulation; these include coal mining and electricity transmission and distribution. These additional industries benefit from the free allowances by selling their allowances to the industries that are points of regulation.

The cases of 100 percent auctioning and of profit-preserving free allocation yield what we termed *net revenue* in earlier chapters, meaning that the revenue from auctioning all or some of the allowances exceeds the adverse revenue impacts from policy-induced reductions in the tax base. In these cases, we consider two alternative ways of achieving revenue neutrality: (1) lump-sum tax rebates to households and (2) reductions in the rates of tax on individual capital and labor income. In contrast, the 100 percent free allocation case does not yield positive net revenue: there is no auction revenue to offset the adverse impact on the tax base.[6] In this case, revenue neutrality is achieved through (modest) increases in individual income tax rates.

Table 7.1 displays, for the years 2020 and 2035, the levels of profit in the reference case and the percentage changes in profit in the two limiting policy cases of 100 percent auctioning and 100 percent free allocation. In percentage terms, the adverse impacts are largest in the energy-intensive industries, with the most significant impacts in the coal-fired electricity generation, coal mining, and natural gas extraction industries.

The difference in impacts between the auctioning and free allocation cases is striking. For the ten industries with the largest percentage losses of profit under 100 percent auctioning, the profit impact is positive under free allocation. The rents generated from free allocation more than compensate for the gross loss of producer surplus imposed by the cap-and-trade program. For any given method of allowance allocation, profits are generally higher when recycling is via individual income tax as opposed to lump-sum rebates. However, the method of allowance allocation has a much larger impact on profits than the recycling method does.

Table 7.2 offers another measure of profit impacts—the percentage change in the present value of profits over the infinite horizon. The

TABLE 7.1 Profit Impacts of Cap and Trade in 2020 and 2035 Under Alternative Allowance Allocation Methods

INDUSTRY	REFERENCE CASE[a]		100% AUCTIONING				100% FREE ALLOCATION[b] INDIVIDUAL INCOME TAX INCREASES	
			2020		2035			
	2020	2035	LUMP-SUM REBATES	INDIVIDUAL INCOME TAX CUTS	LUMP-SUM REBATES	INDIVIDUAL INCOME TAX CUTS	2020	2035
Oil extraction	90.50	58.80	-0.6%	-0.7%	-0.1%	-0.1%	-0.6%	-0.2%
Natural gas extraction	55.92	65.37	-12.8%	-12.8%	-17.6%	-17.5%	58.0%	76.5%
Coal mining	8.89	9.63	-39.7%	-39.6%	-44.0%	-44.0%	73.7%	118.7%
Electric transmission and distribution	42.54	46.53	-10.0%	-9.7%	-8.1%	-7.8%	11.4%	22.4%
Coal-fired electricity generation	12.98	11.70	-61.9%	-61.9%	-78.1%	-78.4%	92.2%	187.7%
Other-fossil electricity generation	7.81	8.96	7.2%	7.7%	-11.2%	-11.1%	61.0%	61.8%
Nonfossil electricity generation	16.28	19.33	24.4%	24.9%	55.2%	56.4%	24.3%	54.9%
Natural gas distribution	19.98	21.36	-5.6%	-5.3%	-8.0%	-7.8%	16.4%	24.1%
Petroleum refining	18.91	20.45	-4.7%	-4.6%	-5.9%	-5.8%	11.6%	17.5%
Pipeline transportation	12.68	14.40	-4.1%	-3.9%	-5.8%	-5.6%	19.6%	26.6%
Mining support activities	19.61	24.23	-5.0%	-4.2%	-3.9%	-3.3%	15.4%	21.9%
Railroad transportation	31.22	39.40	-3.5%	-3.3%	-4.1%	-3.9%	8.4%	10.5%
Construction	49.10	61.10	-1.4%	-0.9%	-1.7%	-1.3%	-1.6%	-1.9%
Manufacturing[c]	117.05	149.69	-1.2%	-1.0%	-1.4%	-1.2%	-1.3%	-1.5%
Transportation[d]	21.76	27.80	-1.2%	-1.0%	-1.2%	-1.1%	-1.3%	-1.3%
Other industries[e]	730.71	920.60	-0.2%	0.1%	-0.2%	0.0%	-0.3%	-0.3%

[a] In billions of 2013 dollars.

[b] Allowances allocated to the ten industries with the largest percentage losses of profit.

[c] Wood products; nonmetallic mineral products; primary metals; fabricated metal products; machinery and miscellaneous manufacturing; motor vehicle; food and beverage; textile, apparel, leather; paper and printing; chemicals, plastics; and rubber.

[d] Air; railroad; water; truck; transit and ground passenger; other and warehousing.

[e] Farms, forests, fishing; other mining; water utilities; trade; communications; owner-occupied housing and real estate.

TABLE 7.2 Profit Impacts of Cap and Trade over the Infinite Horizon[a] Under Alternative Allowance Allocation Methods

INDUSTRY	100% AUCTIONING		PROFIT-PRESERVING ALLOCATION		100% FREE ALLOCATION[b]	
	(1) LUMP-SUM REBATES	(2) INDIVIDUAL INCOME TAX CUTS	(3) LUMP-SUM REBATES	(4) INDIVIDUAL INCOME TAX CUTS	(5) LUMP-SUM REBATES	(6) INDIVIDUAL INCOME TAX INCREASES
Oil extraction	-0.1	-0.1	-0.1	-0.1	-0.1	-0.1
Natural gas extraction	-23.5	-23.5	0.0 (9.0)	0.0 (8.9)	78.8	78.5 (39.0)
Coal mining	-45.9	-45.9	0.0 (2.3)	0.0 (2.3)	153.9	153.2 (10.0)
Electric transmission and distribution	-7.9	-7.6	0.0 (2.1)	0.0 (2.0)	26.3	26.2 (9.0)
Coal-fired electricity generation	-74.7	-74.9	0.0 (4.5)	0.0 (4.5)	250.4	249.4 (19.8)
Other-fossil electricity generation	-18.5	-18.6	0.0 (1.0)	0.0 (0.9)	62.2	61.9 (4.2)
Nonfossil electricity generation	62.7	63.9	62.7	63.5	62.7	62.3
Natural gas distribution	-8.4	-8.2	0.0 (1.0)	0.0 (1.0)	28.1	27.9 (4.4)
Petroleum refining	-6.3	-6.2	0.0 (0.7)	0.0 (0.7)	21.1	21.0 (3.0)
Pipeline transportation	-7.2	-7.0	0.0 (0.7)	0.0 (0.7)	24.3	24.1 (3.0)
Mining support activities	-5.5	-4.9	0.0 (0.9)	0.0 (0.8)	18.3	18.1 (4.0)
Railroad transportation	-3.6	-3.5	0.0 (0.8)	0.0 (0.8)	12.1	12.0 (3.7)
Construction	-2.3	-1.8	-2.3	-2.0	-2.3	-2.4
Manufacturing[c]	-2.1	-1.9	-2.1	-1.9	-2.1	-2.2
Transportation[d]	-2.1	-1.9	-2.1	-2.0	-2.1	-2.2
Other industries[e]	-1.2	-1.0	-1.2	-1.1	-1.2	-1.3
All industries	-1.3	-1.1	-1.2 (23.0)	-1.1 (22.6)	-1.1	-1.2 (100.0)
GDP costs	0.64	0.45	0.64	0.51	0.64	0.69
EV/ton	$42.12	$31.59	$42.12	$340.71	$42.12	$45.32

[a] Figures indicate the percentage changes in the present value of profits over the infinite horizon. Numbers in parentheses are the percent of total allowances allocated to each industry.
[b] Allowances allocated to ten most impacted industries by profit loss.
[c] Wood products; nonmetallic mineral products; primary metals; fabricated metal products; machinery and miscellaneous manufacturing; motor vehicle; food and beverage; textile, apparel, leather; paper and printing; chemicals, plastics, and rubber.
[d] Air; water; truck; transit and ground passenger; other and warehousing.
[e] Farms, forests, fishing; other mining; water utilities; trade; communications; services; owner-occupied housing and real estate.

limiting cases of 100 percent auctioning and 100 percent free allocation are displayed in the first two and last two columns, respectively. In these limiting cases, the distribution of profit impacts across industries is similar to the distribution shown for selected years in the corresponding cases in table 7.1. The middle columns of table 7.2 indicate the needed free allowances in our profit-preserving case, for the ten industries that would experience the largest percentage loss of profit under 100 percent auctioning. The natural gas extraction, coal-fired generation, and coal mining industries would need the largest number of free allowances, requiring about 9, 5, and 2 percent of the total supply of allowances, respectively.

Notably, the free allowances needed by the ten most vulnerable industries together total less than a quarter of the cap-and-trade system allowances. From table 7.2, 23 percent could be given out free, and the remaining 77 percent auctioned. This is so because, according to the production and demand elements in the model, industries are capable of shifting much of the policy costs onto consumers: supply elasticities are large relative to demand elasticities.[7] This limits the amount of free allocation needed to preserve profits.

The bottom rows of table 7.2 compare the GDP and welfare costs of the alternative policies. The three policies yield the same GDP and welfare costs when the revenues are recycled in lump-sum fashion (columns 1, 3, and 5). However, the costs are quite different when recycling involves changes in individual income tax rates (columns 2, 4, and 6). Among these cases, the 100 percent auctioning case is the least costly, since it raises the most revenue and thus leads to the largest cuts in marginal tax rates. The net revenue in this case is sufficient to reduce marginal tax rates on individual labor and capital income by about 1 percent, which corresponds to a 0.26–0.32 percentage point reduction in labor, interest income, dividend income, and capital gains taxes. Under profit-preserving allocation (column 4), the GDP and welfare costs are about 12 and 10 percent higher, respectively, than under 100 percent auctioning. Since some of the allowances are given out free in this case, the policy yields less revenue from auctioning and thus does not allow for marginal rate cuts as large. The costs are highest under 100 percent free allocation (column 6), since in this case there is no auction revenue and thus no opportunity to finance cuts in marginal rates.

The profit-preserving allocations could engender political support from industries receiving free allowances, industries that could otherwise oppose the implementation of an economywide climate policy. However, when recycling takes the form of individual income tax cuts, the profit-preserving case is more costly to the overall economy, since auction revenues are lower and individual income tax rates cannot be reduced as much. Some policy makers could wonder whether profit preservation is worth this additional cost. Still, the profit-preserving case is considerably more cost-effective than 100 percent free allocation, which has GDP and welfare costs over 53 and 44 percent higher, respectively, than under 100 percent auctioning with individual income tax recycling.[8]

The finding that relatively few allowances are needed to preserve profits is consistent with earlier findings by Bovenberg and Goulder (2001), Smith and Ross (2002), and Smith, Ross, and Montgomery (2002). These studies considered only upstream cap-and-trade systems and concentrated on preserving profits in the coal mining and oil and gas extraction industries. Bovenberg and Goulder found that freely allocating about 10 percent of the allowances sufficed to preserve profits in those industries. Smith and Ross found that 9–21 percent would be sufficient to preserve these industries' profits. In a subsequent study, Smith, Ross, and Montgomery (2002) obtained roughly similar results. Goulder, Hafstead, and Dworsky (2010) found that under a wide range of cap-and-trade programs, less than 15 percent of the allowances needed to be offered free to offset profit losses in the eight most vulnerable industries.[9]

Figures 7.2a and b offer a further look at the costs of preserving profits, displaying how these costs change as a function of the stringency of the overall cap and the scope of industries to be offered relief through free allocation. In the low- and high-stringency cases, the overall caps are set so as to yield allowance prices 50 percent higher and lower, respectively, than in the central case. Thus, the allowance prices start in 2019 at $10 and $30 per ton (after a 3-year phase-in), respectively, increase (as in the central case) at an annual rate of 4 percent, and are capped at $30 and $90, respectively, in 2047.

Figure 7.2a displays the additional welfare cost, in terms of the equivalent variation (EV) per ton, of expanding the group of industries receiving profit-preserving allocations. The labels on the horizontal axis identify the

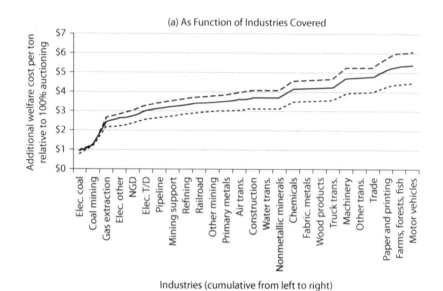

(a) As Function of Industries Covered

Additional welfare cost per ton relative to 100% auctioning

Industries labels (left to right): Elec. coal, Coal mining, Gas extraction, Elec. other, NGD, Elec. T/D, Pipeline, Mining support, Refining, Railroad, Other mining, Primary metals, Air trans., Construction, Water trans., Nonmetallic minerals, Chemicals, Fabric. metals, Wood products, Truck trans., Machinery, Other trans., Trade, Paper and printing, Farms, forests, fish, Motor vehicles

Industries (cumulative from left to right)
receiving profit-preserving free allowances

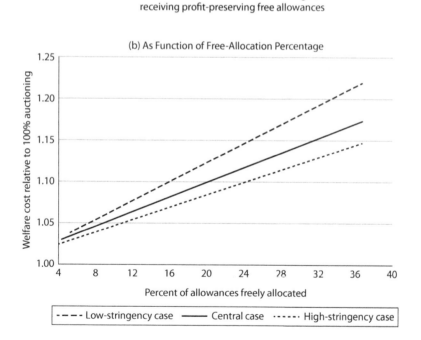

(b) As Function of Free-Allocation Percentage

Welfare cost relative to 100% auctioning

Percent of allowances freely allocated

- - - - Low-stringency case —— Central case ······ High-stringency case

FIGURE 7.2

Welfare Costs of Free Allowance Allocation

added industry. Expanding the scope of profit preservation increases the cost of the policy because it reduces the amount of revenue that can be used to finance individual income tax cuts. As the wobbliness of the figure indicates, the cost implied by adding one industry varies. The added cost depends on the number of allowances the given industry needs to preserve profits. Some industries may have high percentage reductions in profits, but if they have small profit levels to begin with, then the additional cost of profit preservation will be small. Alternatively, some industries may have small percentage changes in profits, but if they are especially large and the reference case volume of profits is high, they will require a large number of free allowances.

Figure 7.2b considers the same profit-preserving efforts as figure 7.2a, this time showing how the relative costs of free allocation increase as the percentage of freely allocated allowances rises as profits are preserved in an increasing number of industries. The figure shows that for each level of stringency, costs increase roughly linearly with the percentage of allowances given out free. As more and more allowances are given out free, the revenue from auctioned allowances falls, but the increase in marginal income tax rates necessitated by this revenue loss is not large enough to raise significantly the marginal excess burden from these taxes. Hence the marginal cost to the economy remains roughly constant.[10]

Figure 7.3a indicates the total number of allowances that the ten most vulnerable industries would receive under profit-preserving free allocation in 2030, as a function of the stringency of the cap-and-trade system. Figure 7.3b displays the value of those allowances in the year 2030 in 2013 dollars. For each industry, increased stringency necessitates a higher value of allowances to maintain profits, but the number of allowances received decreases as higher stringency gives rise to higher allowance prices. The latter effect dominates; hence for all industries, the number of allowances received by industries under profit-preserving free allocation declines with stringency.

Distribution of Profit Impacts Under a Carbon Tax

As discussed earlier, a carbon tax and equivalently scaled cap-and-trade system with 100 percent auctioning produce the same economic outcomes in the E3 model. By "equivalently scaled" we mean that the caps are set

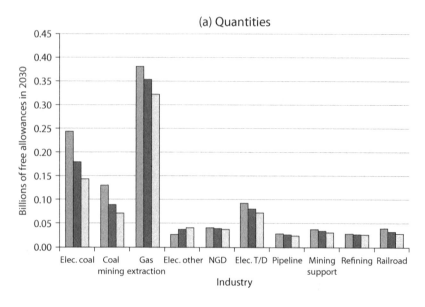

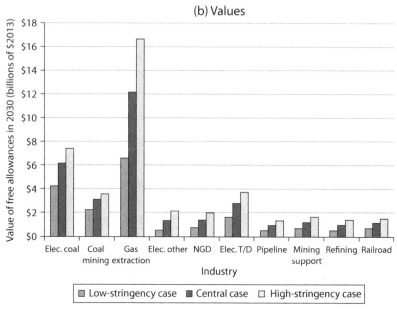

FIGURE 7.3

Quantities and Values of Free Allowances in 2030 by Policy Stringency

to match the levels of emissions under the carbon tax. The cap-and-trade system described earlier in this chapter involved caps corresponding to emissions levels under the central case carbon tax with revenue recycling through lump-sum rebates. Hence the results shown for cap and trade in the first numbered column of table 7.2 are also the impacts for the central case carbon tax with lump-sum recycling.

Table 7.3 reveals the distributional impacts under several other carbon tax policy specifications. The columns numbered 1 and 2 indicate the impacts when revenues are recycled through cuts in individual income tax rates or in the corporate income tax. In keeping with the fact that the marginal excess burden from corporate income tax exceeds that of individual income taxes, recycling through cuts in the corporate income tax rate offers a cost advantage over recycling through cuts in the individual income tax. The profit losses to the most vulnerable industries are slightly smaller under corporate income tax recycling, but the distribution of these losses across industries is similar. The natural gas extraction industry, for example, experiences a profit loss of about 23 percent under individual income tax recycling, but it experiences a smaller profit loss of about 20 percent under corporate tax recycling. Notably, several industries that experience profit losses under individual income tax recycling enjoy increases in profits under corporate income tax recycling.

Altering the Distribution of Industry Impacts

How can a carbon tax be designed to achieve more even profit impacts across industries, or avoid significant losses of profit to especially vulnerable industries? One way is to provide vulnerable industries with exemptions to the carbon tax for some of their emissions. Under this approach, firms would be taxed on only the emissions exceeding the level given by the exemption. Firms would still face the tax at the margin; thus their incentives to reduce emissions would be the same as in the case of a carbon tax with no exemptions. The effects on output prices would also be the same.

The incentives in this case are also the same as under an equivalently scaled cap-and-trade program in which the allocation of free allowances corresponds to the allocation of emissions exemptions under the carbon tax. And because the incentive effects are the same, a carbon tax with

TABLE 7.3 Carbon Tax with Compensation to Industry: Impacts on Profits, GDP and Welfare
Percentage Changes in the Present Value of Profits over the Infinite Horizon

PERCENTAGE CHANGE IN PROFITS BY INDUSTRY[a]	NO TARGETED COMPENSATION		TARGETED COMPENSATION TO TEN MOST VULNERABLE INDUSTRIES		
	(1) INDIVIDUAL INCOME TAX CUTS	(2) CORPORATE INCOME TAX CUTS	(3) TRADABLE TAX EXEMPTIONS[a]	(4) CORPORATE INCOME TAX CREDIT[b]	(5) CORPORATE INCOME TAX CUT[c]
Oil extraction	-0.1	6.8	-0.1	-0.1	-0.4 (0.40)
Natural gas extraction	-23.3	-20.3	0.0 (8.9)	0.0 (10.6)	0.0 (0.03)
Coal mining	-45.7	-45.7	0.0 (2.3)	0.0 (2.7)	-30.8 (0.00)
Electric transmission and distribution	-7.6	-5.5	0.0 (2.0)	0.0 (2.4)	0.0 (0.31)
Coal-fired electricity generation	-74.6	-75.0	0.0 (4.5)	0.0 (5.4)	-59.2 (0.00)
Other-fossil electricity generation	-18.3	-14.8	0.0 (0.9)	0.0 (1.1)	0.0 (0.34)
Nonfossil electricity generation	63.4	66.1	63.1	63.1	48.3 (0.40)
Natural gas distribution	-8.1	-5.7	0.0 (1.0)	0.0 (1.2)	0.0 (0.34)
Petroleum refining	-6.1	-3.2	0.0 (0.7)	0.0 (0.8)	0.0 (0.29)
Pipeline transportation	-7.0	-3.3	0.0 (0.7)	0.0 (0.8)	0.0 (0.34)
Mining support activities	-4.9	-0.5	0.0 (0.8)	0.0 (1.0)	0.0 (0.34)
Railroad transportation	-3.4	0.3	0.0 (0.8)	0.0 (0.1)	0.0 (0.37)
Construction	-1.8	0.5	-1.9	-2.0	-1.0 (0.40)
Manufacturing[d,g]	-1.8	0.6	-1.9	-1.9	-1.6 (0.40)
Transportation[e,g]	-1.9	0.2	-2.0	-2.0	-1.6 (0.40)
Other industries[f,g]	-1.0	0.2	-1.0	-1.0	-0.8 (0.35)
All industries[g]	-1.1	0.2	-1.1 (22.7)	-1.1 (26.9)	-0.8 (0.36)
GDP costs[i]	0.45	0.19	0.50	0.51	0.26
EV/Ton	$31.30	$13.95	$34.52	$34.73	$23.71

Note: In targeted compensation scenarios, revenues not devoted to compensation are recycled through cuts in individual income tax rates.

[a] Numbers in parentheses reflect the required number of tradable exemptions as a percent of annual carbon tax base.

[b] Numbers in parentheses are the initial values of corporate income tax credit in 2013 billion dollars. Credit grows at expected nominal GDP growth annually.

[c] Numbers in parentheses are levels of industry-specific corporate tax.

[d] Percentage change in present value of profits over the infinite horizon.

[e] Wood products; nonmetallic mineral products; primary metals; fabricated metal products; machinery and miscellaneous manufacturing; motor vehicle; food and beverage; textile, apparel, leather; paper and printing; chemicals, plastics, and rubber.

[f] Air; water; truck; transit and ground passenger; other and warehousing.

[g] Farms, forests, fishing; other mining; water utilities; trade; communications; services; owner-occupied housing and real estate.

[h] Weighted average, using 2013 output shares.

[i] Percentage change in present value of GDP, 2017-2050.

tradable exemptions produces the same economic outcomes as the equivalent cap-and-trade program with freely allocated allowances.

By lowering firms' tax burdens without implying lower output prices, the exemptions cause firms to earn rents much in the way that free allowances yield rents under cap and trade. Firms in industries that are not points of regulation—firms in these industries do not face the carbon tax, but nonetheless would feel its impact—can benefit by selling exemptions to firms in the industries that are points of regulation. This parallels the way that, under cap and trade, firms in uncovered industries can benefit by selling the allowances they have received to firms in covered industries.

As with free allowances, tradable exemptions reduce the amount of revenue generated by the tax. In the policy simulation for the third numbered column of table 7.3, tradable exemptions are distributed in the same way that allowances were distributed free to preserve profits under cap and trade earlier: just enough tradable exemptions are given to the ten most affected industries to preserve the present value of their profits. The revenues from the policy are used to finance cuts in individual income tax rates. The numbers in parentheses are the required number of exemptions as a percentage of the annual carbon tax base. The natural gas extraction, coal-fired electricity generation, and coal mining industries receive the most exemptions, just as they received the most allowances in the cap-and-trade policy considered earlier.[11] Compensation through tradable exemptions comes at a cost, since the exemptions imply lower carbon tax revenue and less potential to reduce preexisting taxes. The welfare cost of the tradable exemption policy is approximately $35 per ton, about 10 percent higher than the cost under the zero compensation policy with individual income tax cuts (column 1).

Targeted (that is, industry-specific) corporate tax credits provide another way to alter the distribution of profit impacts. Column 4 of table 7.3 displays the results when such credits are offered to the ten most vulnerable industries. In this policy, we provide these industries with credits each year that can be used to reduce (or eliminate) the corporate tax burden. In each of these industries, the nominal value of the credit is specified as growing at approximately the growth rate of nominal GDP, and the number of credits along this time path is just large enough to prevent the reduction in the present value of profits that the industry would otherwise experience. The numbers in parentheses indicate the initial value of the credit received by each industry in 2016 (in billions of 2013 dollars). For a few industries (e.g., coal mining and coal-fired generation), the value of the credit is larger than the annual corporate tax payment of the industry in the reference case.[12] The efficiency cost, assuming that the net carbon revenue is first used to finance the credits, with remaining revenue devoted to cuts in individual income taxes, is approximately equal to the cost of profit-preserving tradable exemptions: both policies raise the welfare cost by approximately $3 per ton reduced, relative to the most cost-effective policy (in column 1).

Column 5 of table 7.3 shows the final industry compensation policy considered: industry-specific cuts in corporate tax rates, with the rate cuts scaled to prevent a reduction in the present value of profits.[13] The numbers in parentheses indicate the level of the industry-specific corporate tax rates after the rate reductions. For some industries—notably natural gas extraction, coal mining, and coal-fired generation—entirely eliminating the corporate tax is not sufficient to prevent a profit loss. This policy is the most cost-effective of the three compensation policies considered. In fact, it is less costly than the policy in column 1 that involves no compensation and devotes all the revenue to cuts in individual income tax rates: the GDP and welfare costs in column 5 are 41 and 24 percent lower, respectively, than those in the first column.[14] However, the GDP and welfare costs are 35 and 70 percent higher than the respective costs displayed in column 2, which involves equal corporate income tax reductions in all industries.

Regional Impacts of Emissions Pricing

Because states differ in their industrial composition, the impact of emissions pricing will vary at the state level. Figures 7.4a, b, and c reveal some differing impacts on state-level profits. Figures 7.4a and b consider impacts of the carbon tax policy with recycling via economywide individual income tax cuts (7.4a) and corporate income tax cuts (7.4b). Figure 7.4c displays the profit impacts with targeted industry compensation through tradable tax exemptions. For each state, the percentage change is a weighted average of the profit impacts in columns 1 (individual income tax cuts), 2 (corporate income tax cuts), and 3 (tradable exemptions) of table 7.3. The weights are calculated by using data on the total value of shipments by industry by state from the 2012 Economic Census (2015).[15]

(a) Cuts in individual income taxes

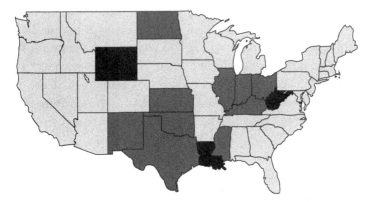

(b) Cuts in corporate income taxes

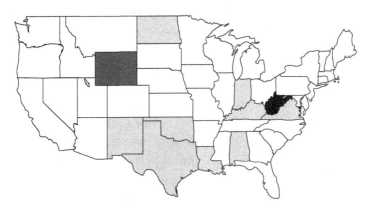

(c) Tradable tax exemptions

☐ 0% to 1%
☐ −2% to −0.1%
■ −4% to −2.1%
■ −11% to −4.1%

FIGURE 7.4

Percentage Change in the Present Value of Profits, by State

Under individual income tax cuts, no states show average profit increases. The profit losses are especially high in states with large coal or natural gas resources such as Wyoming, West Virginia, North Dakota, Kentucky, Louisiana, Oklahoma, and Texas. Under corporate income tax cuts, average profit impacts are positive in some states and negative in others, and more states enjoy increases in profits than endure profit losses. All the "high-impact states" mentioned earlier still experience profit losses.

In the targeted compensation case, the high-impact states tend to have smaller profit losses than the average across all states. This reflects the free allowances that are being used to prevent profit losses in energy-intensive industries within these states. Among the high-impact states, the states with undiversified economies (West Virginia, Wyoming, Kentucky, and Louisiana) have the smallest profit losses under this policy.

It should be noted that the distribution across states in average profit impacts will differ from the distribution of average welfare impacts of the residents of various states. Other things equal, the largest adverse economic impacts will be experienced by individuals who disproportionately own capital in the firms in industries that suffer the largest negative profit impacts. Some of these individuals will reside in states that do not experience relatively large profit losses.

Summary

The emissions pricing policies we have examined produce significant adverse profit impacts on industries that intensively supply or use carbon-based fuels. However, as we've shown here, the potential negative profit effects can be mitigated through free allowance allocation under a cap-and-trade program or through targeted compensation via revenues from a carbon tax. Preventing profit losses in the ten most impacted industries, through either free allowances under cap and trade or tradable tax exemptions under a carbon tax, raises welfare costs by about 10 percent when remaining revenues are recycled through cuts in the rates of individual income tax rates.

Other ways to preserve profits are to provide corporate tax credits or corporate income tax cuts for firms in the most vulnerable industries. Relative to a policy with individual income tax cuts and no compensation, providing corporate tax credits raises welfare costs by about 11 percent;

offering industry-specific corporate tax rate reductions *lowers* welfare costs by about 24 percent. (In both cases, remaining revenues are recycled via individual income tax cuts.) The latter case includes some corporate tax recycling, whereas the no-compensation case only employed individual income tax recycling. This implies lower costs because it includes cuts in a more distortionary tax—the corporate tax. Thus, preventing losses of profits either involves a small increase in cost or reduces overall policy costs in the case of policies with individual income recycling. Some policy makers might find such compensation attractive, given that the additional costs are either small or negative relative to the costs in the absence of compensation.

IMPACTS ACROSS HOUSEHOLD GROUPS

In considering the impacts of various climate-change policies, Chapters 5 and 6 focused on the economywide costs, employing the *equivalent variation* (EV), as the key cost-measure. The EV indicates the dollar-equivalent to the change in utility experienced by the E3 model's representative household. Although this representative impact is useful information, the welfare impacts can differ significantly across household groups, and the distribution of impacts across households is highly relevant to fairness and political feasibility. Indeed, the household distribution can be considered to be even more important than the distribution across industries, since costs of climate policies are ultimately felt by people.

Here we consider how the welfare impacts are distributed across household groups. For each household group, we assess the overall welfare (EV) impact and the underlying use-side and source-side impacts that contribute to the overall welfare outcome. Use-side impacts are the changes in welfare caused by changes in the prices of goods and services that households purchase. Source-side impacts are the changes in welfare caused by changes in the value of the household's labor, capital, and transfer endowments.

We assess the impacts across households under a range of policy designs, including policies with and without compensation schemes that alter the distribution of welfare impacts. We also show how the overall economywide costs are affected by the type of compensation that might be offered.

The Disaggregated Household Model

To consider the effects for different household groups, we link the E3 model with one that distinguishes five population groups—income quintiles. For both the reference case and each climate policy considered, the E3 model generates time paths of income (from labor, capital, and transfers) and of prices of goods, services, and leisure. Taking the returns to factors and the prices of goods, services, and leisure from the E3 model, the disaggregated household (DH) model determines, for each quintile, the consumption, labor supply, and savings choices that maximize utility over time.

Climate policies can have very different impacts across household groups because households differ both in their relative reliance on labor, capital, and transfer income (implying differing source-side impacts) as well as in their patterns of expenditure on goods and services (implying differing use-side impacts). The DH model recognizes these differences, enabling our analysis to account for how a given climate change policy will lead to different choices for consumption, saving, and labor supply across households. Although the household responses differ, for each household the functional form of the utility function is the same as the one for the representative E3 household. Differences in household responses reflect differences in some parameters of the utility functions as well as differences in endowments and expenditure patterns.

A strength of this approach is that in assessing the impacts on households, we are able to account for both the source- and the use-side impacts—both the impacts on labor, capital, and transfer incomes as well as the impacts on the prices of goods and services purchased. We consider, for each quintile, the relative importance of these impacts and their contribution to overall welfare.[16]

Two capabilities distinguish our approach from the approaches of earlier studies. First, by exploiting the multiperiod aspects of the E3 and DH models, we are able to explore how the impacts of policies across household groups change over time. We will see that the distribution pattern differs substantially, depending on the time interval over which the impacts are considered. Second, by combining results from the E3 and DH models, we can evaluate the potential costs of avoiding uneven impacts across household groups through targeted compensation schemes. These potential trade-offs—between aggregate cost and a more even distribution of outcomes—are of considerable interest to policy makers.

In the following three sections we offer some specifics related to the data, structure, and application of the DH model. Details are provided in appendix D.

Data Consistency Procedures and Parameterization

The main data consistency task is to ensure that the benchmark period quintile-level data, when aggregated, match the data for the E3 model's representative household in the same benchmark period. To accomplish this, we first use detailed data on household expenditures to obtain shares of expenditure by quintile for each good (details are provided in appendix D). We then apply these shares to derive quintile-level data that match, in aggregate, the level of expenditure on each good. To achieve consistency in prices, first we assume the representative household in each quintile faces the same prices as those faced by the E3 model's representative household, and then we determine the composite consumption price $\bar{p}^q$ for each quintile q. Since the expenditure shares differ by quintile, these composite prices differ.

Another consistency task is to ensure that the DH model's benchmark endowments of labor, capital, and transfers, when added across the quintiles, match the endowments in the E3 model. We also need to ensure that the budget constraint of the representative household for each quintile is satisfied. In our approach, the factor returns (that is, the after-tax wage $\bar{w}$ and after-tax financial return $\bar{r}$) are assumed to be equal across quintiles; these match the values in the E3 model. For levels of labor, capital, and transfer endowments, we use detailed data (see appendix D) on household income sources to obtain income shares and apply these shares to allocate labor, capital, and transfer endowments from the E3 model's representative household to generate benchmark quintile-level data. Lump-sum taxes for each household are then allocated from the E3 representative household to each quintile household such that the budget constraint is binding for each household.

The next task is parameter calibration. We calibrate parameters of the utility function for the representative household of each quintile to satisfy the condition that the specified benchmark levels of consumption, leisure, and other variables are consistent with utility maximization and satisfy the household's budget constraint. In this step, we assume that each household maintains the same compensated elasticity of labor supply as the E3 representative household, and that each household maintains the same labor-leisure ratio.[17] Finally, each household's preferences over the consumption goods are calibrated to be consistent with our data on expenditures by quintile and goods.

Policy Simulation and Welfare Calculations

With the consistent and fully parameterized dataset, we are able to determine each household's utility-maximizing consumption, labor supply, and savings choices in the reference case and policy cases. We apply to each quintile the E3-model prices for each period of the reference and policy cases. These include the prices for consumption goods and services, the after-tax wage, and the after-tax return to capital. The factor endowments (capital, labor, and transfers) are set to be consistent with the E3 model in each reference and policy case.[18] Then, as described in appendix D, we solve the disaggregated household model for each quintile in the reference and policy cases; each household chooses consumption, leisure, and savings to maximize lifetime utility.[19]

We calculate the impact of the policy change on the welfare of each quintile through the use-side and source-side impacts. As discussed later, we consider and compare outcomes under several variants of the measures of the use- and source-side impacts. Appendix D derives the formulas for the use- and source-side impacts and shows the conditions under which they perfectly total the overall welfare impact.

Data Sources for the DH Model

Here we briefly describe key sources of data for the DH model. Appendix D offers details. We obtain household expenditures on each consumer good by quintile using Consumer Expenditure Survey (CEX) data collected by the U.S. Department of Labor's Bureau of Labor Statistics (BLS) (Bureau of Labor Statistics 2014). This includes detailed 2013 and 2014 microdata as well as less detailed BLS website data. The Consumer Expenditure Survey includes both an Interview Survey and a much more detailed Diary Survey. The Interview Survey focuses on large purchases, such as expenditures on housing, vehicles, and health care. The Diary Survey gives greater focus to small, frequent purchases such as spending on food, beverages, and tobacco. We combine the information from both surveys to obtain a complete listing of expenditure shares.

We organize the data two ways, sorting the households by total expenditure and by total income. Sorting households by current income conveys useful information but does not always convey economic status: some households with low income have substantial wealth from inheritances or prior saving. Households headed by retirees often have this feature. Household expenditure has the potential to offer a better picture of economic status or net worth. For this reason, later we focus mainly on the impacts across households sorted into quintiles by expenditure.

We derive average after-tax income per quintile (ordered by before-tax income) from labor, capital, and transfers from the Congressional Budget Office (CBO) (2013). Table 7.4 displays the average income data from the CBO. For each income source, we calculate the share of total income received by each quintile and use these shares to allocate E3 representative household income across the household groups. For each household quintile, we allocate benchmark financial wealth according to the capital income shares.

Welfare Impacts

Here we first evaluate the use- and source-side impacts individually under our central case economywide carbon tax. Later, we consider the full welfare impacts by combining the use- and source-side impacts.

TABLE 7.4 After-Tax Income by Source and Quintile
Average household income in thousands of 2013 dollars

	QUINTILE (BEFORE-TAX INCOME)				
INCOME SOURCE:	0-20TH	20-40TH	40-60TH	60-80TH	80-100TH
Labor	$13.0	$23.7	$36.4	$57.0	$115.9
Capital	$2.6	$4.5	$9.0	$16.0	$78.4
Transfers	$9.6	$16.2	$16.7	$15.0	$12.0
Total	$25.2	$44.4	$62.1	$88.0	$206.3

Use-Side Impacts

Our analysis accounts for the fact that households tend to substitute away from goods or services whose relative prices have increased and toward those whose relative prices have decreased. (Initial studies of the distributional impacts of carbon and other environmental taxes often assumed fixed expenditure shares. Further, our approach allows for shifts between consumption of goods and services and consumption of leisure.

We gauge the use-side impacts two ways. One approach gives the dollar-equivalent change in utility at a given moment (period) of time. The other measures the welfare impact over an interval of time.

Figure 7.5 displays single-period use-side impacts by quintile and under the four recycling options. In the recycling cases involving tax cuts, we assume that the rate cuts are the same for all quintiles. The households have been ordered by expenditure. Impacts are shown for the years 2020, 2030, and 2050 and are expressed as a percentage of reference case expenditures.

The figure's two columns calculate the impacts in two ways. In the left-hand column, the use-side impact accounts for the policy-induced changes in the prices of goods and services, excluding the impact on the price of leisure (another "good" that a household can "purchase" by working less and sacrificing income). The right-hand column offers results from a broader calculation of the use-side impact, one that accounts for policy-induced changes in the price of leisure.

Figure 7.5 yields four key findings. First, under each of the four recycling options, the use-side impact is regressive: the welfare impact is more negative, the lower the expenditure rank of the quintile. This reflects the fact that lower-quintile households spend a larger share of their incomes on carbon-intensive goods and services than do higher-quintile households. The outcome is regressive regardless of whether changes in the price of leisure are ignored (left-hand column) or considered (right-hand column).[20]

Second, for all quintiles, the magnitude of the use-side impact increases with time, paralleling the increasing size of the carbon tax and the associated increases in the scale of the price impacts.

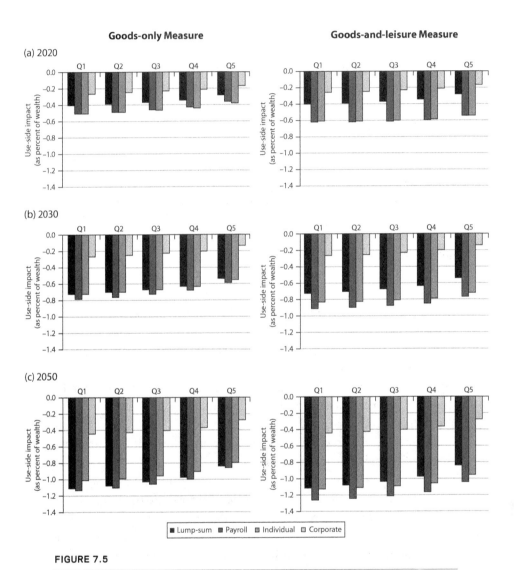

FIGURE 7.5

Use-Side Impacts by Year by Quintile, Selected Years

Third, the size of the use-side welfare impact in any given year depends on the type of recycling. The impacts are smallest when recycling is via cuts in the corporate income tax. This is in keeping with the fact that the corporate tax is the most distortionary, and reducing the rate of this tax produces the largest beneficial revenue-recycling effect.

Fourth, under recycling in the form of payroll tax cuts or individual income tax cuts, the use-side impacts are larger when changes in the price of leisure are accounted for: effects in the right-hand column are larger than those in the left-hand column. Each of these two forms of recycling involves cuts in the tax on wages. This raises the after-tax wage, which is also the price of leisure. Thus, the use-side impacts are larger when leisure's price is considered.

Figure 7.6 displays the use-side impacts for the same years, this time for households ordered by their income. The results are similar to those in figure 7.5, though the results are slightly more regressive. As found in previous studies, ordering households by income yields more regressive impacts because this ordering places low-income retirees in low quintiles. Retirees tend to spend a large fraction of income on carbon-intensive goods.

Figure 7.7 shows the use-side impacts when measured over time intervals rather than points in time, indicating the effects over the intervals 2017–2020 and 2017–2040, as well as over the interval of infinite length that begins in the year 2017.[21] As with our first measure, it provides the dollar equivalent to the change in utility. And as before, the two columns compare results without and with consideration of impacts on the price of leisure. In this figure and in the remaining figures in this chapter, the results are for households ranked by expenditure.

The results in figure 7.7 parallel those in figure 7.5. The results are again regressive and increase with the amount of attention to the longer term. And accounting for the impact on the price of leisure again expands the adverse welfare impact in the cases of recycling via cuts in the payroll tax or individual income tax.

Source-Side Impacts

The regressivity of the use-side impacts is a robust result. But it would be premature to conclude from the evidence presented thus far that the

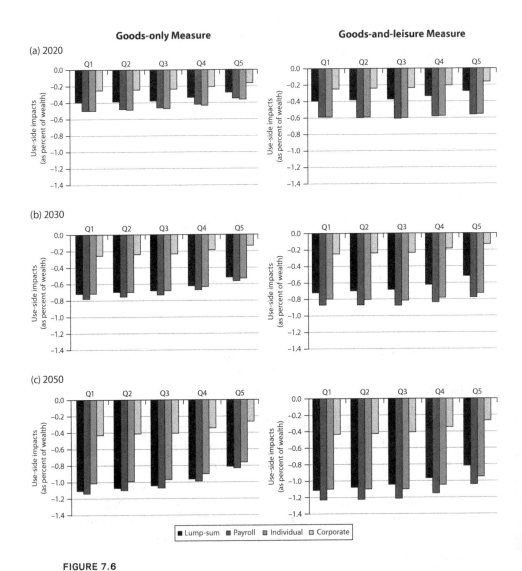

FIGURE 7.6

Use-Side Impacts by Year Quintile, Selected Years, Households Ordered by Income

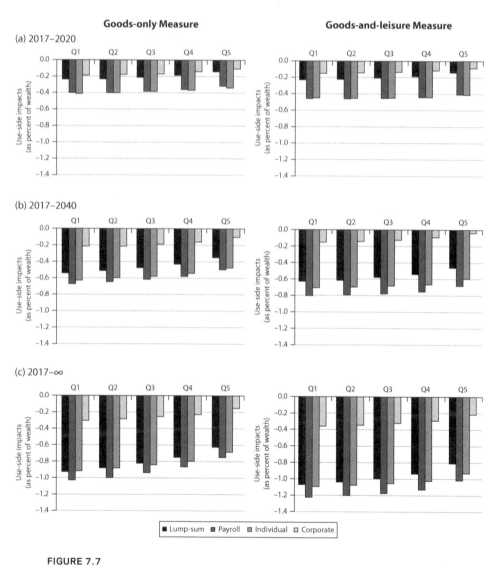

FIGURE 7.7

Use-Side Impacts by Quintile, Selected Intervals

carbon tax policies we have considered will be regressive. The overall effect depends as well on the source-side impact. We will see that the source-side impacts are generally progressive and change the picture dramatically.

Figures 7.8 and 7.9 display the impacts on the source side. We again consider the impacts at given points in time (figure 7.8) and over specified intervals of time (figure 7.9), and under the four forms of revenue recycling considered previously. Under the policies with recycling via lump-sum rebate policies, the source-side effects for each quintile will depend on a policy choice: namely, the specified share of total rebates to be received by each quintile. In figures 7.8 and 7.9, the results for the case of lump-sum rebates are from policies in which each quintile receives one-fifth of the total rebate provided in each period. Later in the chapter, we will consider alternative rebate schemes aimed at achieving certain objectives in terms of the distribution of impacts across households.

The two columns in each figure show the results under two measures of the source-side impact. The left-hand column employs a narrower, "income-only" measure of the source-side impact, one that considers only the policy's effects on after-tax labor income, after-tax capital income, and transfer income. The right-hand column offers a measure that is broader in two ways. First, in connection with labor, it considers the impact of policy on each household's overall endowment of labor—which is the sum of the value of labor supplied and the value of the household's nonlabor (leisure) time. When a household decides to work less, this reduces the labor income that the household receives. Measuring the welfare impact would overstate the welfare loss associated with this change, since the reduction in income is compensated for by the value of the increase in nonwork (leisure) time. Our broader measure accounts for this offset. Second, the broader measure accounts for the impact of policy on each household's savings in a given period or during the time interval of focus. Any increase (decrease) in saving implies greater (lower) potential for future consumption and utility. Although some of this change in future consumption may occur beyond the period or time interval of focus, the source of this change is in the period or during the interval of focus; hence it can be attributed to those points in time. Accounting for the savings impact also has the virtue of enabling the sum of the source- and use-side

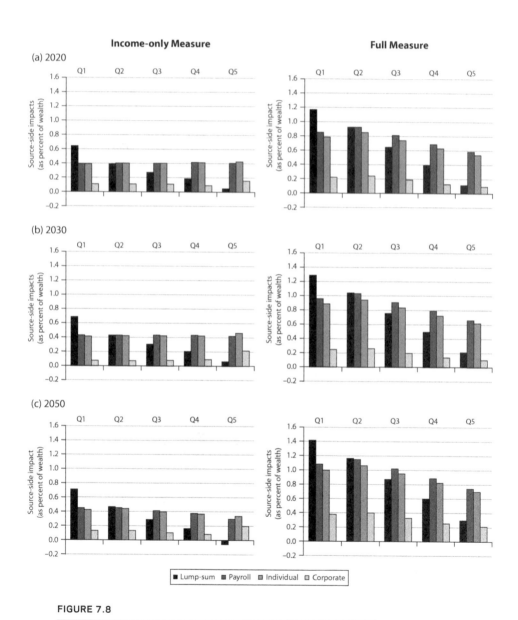

FIGURE 7.8

Source-Side Impacts by Quintile, Selected Years

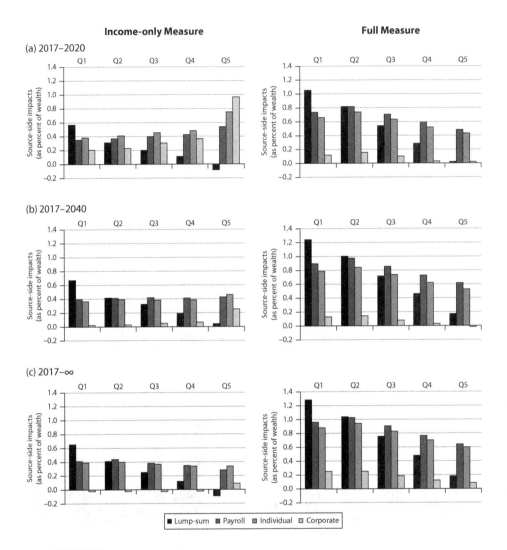

Income-only Measure **Full Measure**

(a) 2017–2020

(b) 2017–2040

(c) 2017–∞

■ Lump-sum ▨ Payroll ▤ Individual ▢ Corporate

FIGURE 7.9

Source-Side Impacts by Quintile, Selected Intervals

impacts to match perfectly the overall welfare impact, as measured by the equivalent variation for the period or interval in question.

It is important to keep in mind that the source-side impacts reflect changes in nominal income or the nominal value of endowments. The use-side impacts account for how changes in prices alter the real purchasing power associated with the nominal changes in income or endowment value.

The key messages from figures 7.8 and 7.9 are similar. First, the source-side impacts are positive, in contrast with the impacts on the use side. One factor behind the positive welfare impacts is revenue recycling. Each form of recycling contributes to nominal income: the lump-sum rebates do so directly, while the cuts in the marginal rates of payroll, individual income, or corporate income taxes do so by increasing the after-tax returns to factors. The positive impact also is due in part to changes in nominal transfers. As indicated earlier, the time path of real transfers under the carbon tax policy is set to match the reference case time path. To the extent that the carbon tax raises the price level, nominal transfers must be increased to maintain the same real value.

Second, the progressivity of the source-side impacts depends on the method of revenue recycling. The progressive outcome is strongest in the case of recycling through lump-sum rebates, in keeping with the fact that the rebates (of equal value for every household) are larger relative to the household's benchmark expenditure, the lower the quintile (or benchmark expenditure) of the household. Also contributing to progressivity is the fact that nominal transfers increase under the carbon tax, as described above, and households in the lower quintiles have a higher share of income from government transfers than do households in higher quintiles. Further, the carbon tax tends to reduce returns to capital more than to labor. Because higher quintiles rely more on capital income than do lower quintiles, this exerts a progressive impact. Under the income-only measure, source-side impacts are proportional in the case of recycling via payroll tax cuts and slightly regressive in the case of individual income tax cuts. These impacts are strongly regressive under corporate income tax cuts, as the upper quintiles benefit from reductions in capital taxes.

Third, the source-side impacts differ considerably when the broader measure is employed. First, the impacts are larger in magnitude. Recycling through cuts in the payroll tax or the individual income tax reduces labor taxes and thereby raises the after-tax wage. This not only increases labor income but also raises the value of leisure. The broader measure captures this latter effect by considering the impact on the labor time endowment. Hence the source-side impact is larger. Under lump-sum rebates and corporate income tax cuts, the impacts also are generally larger, reflecting the inclusion of the impact of savings on current consumption under the full measure of source-side impacts. Finally, in contrast with the outcome under the income-only measure, the source-side impact is progressive under all forms of recycling, though the magnitude of progressivity varies by the method of recycling.[22] This shows that including the effects on the time endowment and savings is crucial to the measurement of source-side impacts.

Overall Welfare Impacts

The use- and source-side effects combine to produce the overall welfare impact. As mentioned earlier (and derived in appendix D), the EV measure of welfare is exactly equal to the sum of our broader use- and source-side impacts. Figures 7.10 to 7.11 display the overall welfare impacts, as measured by the equivalent variation.

These figures show the impacts across quintiles at given points in time and over given intervals of time, respectively, as a percentage of reference case expenditures. They show that the overall impacts are progressive under recycling via lump-sum rebates: the very progressive source-side impacts outweigh the regressive impacts on the use side.[23] Notably, under our most complete measure of the welfare impacts (the infinite-horizon EV measure in figure 7.11c), the bottom two quintiles are better off under a carbon tax with lump-sum rebates than in the absence of a carbon tax. Under payroll tax recycling, the results are very different, but still generally progressive.[24] This reflects the very different source-side impacts, observed earlier, stemming from the payroll tax cut's positive impact on the after-tax wage. Under individual income tax recycling, the results are similar to, but somewhat smaller in magnitude than, the results under payroll tax recycling. This policy's positive impact on the after-tax wage again works toward the generally progressive outcome. Under corporate income tax recycling, the impacts are smaller than under the other recycling methods, in keeping with the offsetting (and generally smaller) use- and source-side impacts under this recycling method.

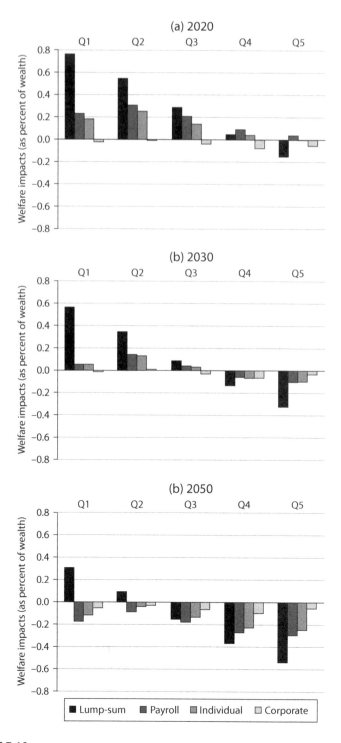

FIGURE 7.10

Overall Welfare Impacts by Quintile, Selected Years

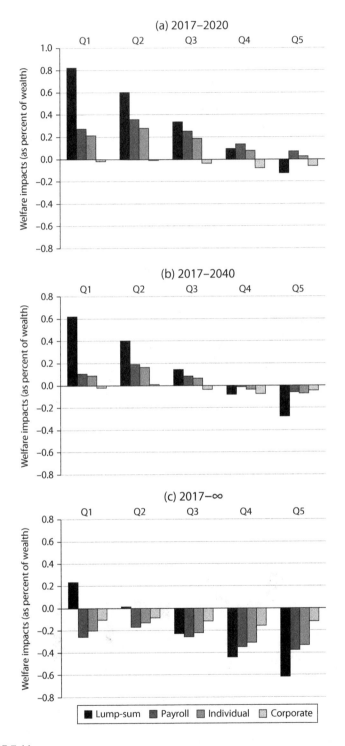

FIGURE 7.11

Overall Welfare Impacts by Quintile, Selected Intervals

Key Findings

It is often claimed that a carbon tax will have a regressive impact, but this conclusion often stems from an exclusive focus on the use-side impacts. A key message from our results is that to gauge the progressivity or regressivity of the carbon tax, it is critical to consider both use- and source-side impacts. We have seen that the source-side impacts tend to be progressive, in many instances fully offsetting the regressivity of the use-side effect.

Our results also show that both the scale and the regressivity or progressivity of the overall (use- plus source-side) impacts depend importantly on the method of recycling. Under recycling via lump-sum rebates, the overall impact is significantly progressive; under corporate income tax recycling, it is nearly proportional; and under payroll tax and individual income tax recycling, it is progressive. The scale of the overall impact is much smaller under corporate income tax recycling than under the other recycling approaches. Although the scale of the impacts varies over the time intervals considered, the progressivity (or proportionality) is robust across the time intervals.

Compensation Options to Offset Uneven Distributional Impacts

From one perspective, our results might seem to have vanquished what had been considered a trade-off between cost-effectiveness and fairness. Under the most cost-effective forms of recycling—cuts in individual or corporate income tax rates—the overall welfare impacts are either proportional or progressive. As a percent of baseline expenditure, the adverse impacts on the lowest two quintiles tend to be no worse than the impacts on the higher quintiles. This could suggest that no additional compensation elements are needed to bring about a fair outcome.

It is worth noting, however, that fairness can connect with the absolute impacts as well as the relative effects. Some interested parties might regard as unfair any policy that causes harm to low-income households. If one adopts this perspective, then a trade-off does in fact emerge between cost-effectiveness and fairness, since (as can be seen from figures 7.10 and 7.11) the most cost-effective forms of recycling produce adverse impacts on the representative household in the lowest two quintiles.

Here we perform additional experiments to quantify this potential trade-off. We examine the impacts of two "hybrid" policies that involve a combination of recycling through lump-sum rebates and recycling through cuts in either individual or corporate income tax rates. Some of the net revenue from the carbon tax is devoted to lump-sum rebates, while the rest is devoted to one of the two tax cuts. Under the two hybrid policies, the rebates are targeted to the lowest two income quintiles and are at a level just sufficient to prevent a welfare loss to the second quintile. (The lowest quintile experiences a small welfare gain under this approach.)

Figure 7.12 shows the distribution of welfare impacts from the two hybrid policies as well as those from the previously discussed "pure" policies involving recycling through lump-sum rebates or tax cuts alone. Under the hybrid policy that includes individual income tax cuts, the first quintile experiences a small welfare gain and, by construction, the second quintile has no welfare impact. (Hence for these policies the bars do not appear on the figure for the second quintile.) For the third quintile, the welfare impact is slightly worse than under the other policies, while for the fourth and fifth quintiles the impacts are between those under lump-sum and individual income tax recycling.

Table 7.5 compares the economywide welfare costs in the hybrid cases with those in the pure recycling cases. Compared with the case where all recycling is through cuts in individual income taxes, introducing lump-sum rebates to provide targeted compensation to the lowest two quintiles (and recycling the remaining net revenue through individual income tax cuts) increases aggregate costs by about 5 percent. Compared with the case where recycling is solely through cuts in corporate income taxes, providing the targeted compensation through rebates (and recycling remaining revenues through corporate tax cuts) raises costs by about 23 percent. We leave it to the reader to assess the importance of the distributional objectives served by these policies and decide whether achieving these objectives is worth the efficiency costs.

Distributional Impacts Under Alternative Climate Policies

Using the same framework as the one applied earlier, we also consider the distributional impacts under alternative policies. Figures 7.13a and b

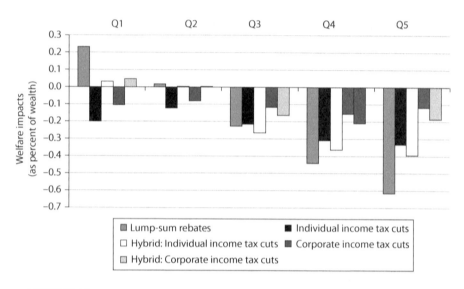

FIGURE 7.12

Welfare Impacts by Quintile: Policies with Targeted Compensation
(Impacts over the Infinite Horizon)

show the overall (use- plus source-side) impacts for the CES policy and the gasoline tax increase. We do not display the impacts under cap and trade; for a cap-and-trade policy with auctioning of emissions allowances, they are the same as under the carbon tax.

Figure 7.13a compares the CES policy's impact with those of an equivalently scaled power-sector-only carbon tax, with recycling of carbon tax revenues through either lump-sum rebates or cuts in individual income tax rates.[25] Note that the axes in the figure differ from those in figures 7.11 and 7.12, to make clear the pattern of the impacts, which are smaller than in the previous figures. The CES policy's impact is close to proportional: slightly regressive use-side impacts are offset by somewhat progressive effects on the source side. The distributional impacts of power-sector-only carbon tax follow a similar pattern to those from the broader central case carbon tax, though the effects are smaller in magnitude. With lump-sum rebating, the outcome is progressive, reflecting the progressive source-side influence of the rebates. With recycling through individual income tax cuts, the outcome is roughly proportional, for reasons similar to those

TABLE 7.5 GDP and Welfare Costs of a Carbon Tax
Under Alternative Recycling Methods

	RECYCLING METHOD				
	LUMP-SUM REBATES	CUTS IN INDIVIDUAL INCOME TAXES	CUTS IN CORPORATE INCOME TAXES	HYBRID: REBATES AND CUTS IN INDIVIDUAL INCOME TAXES[b]	HYBRID: REBATES AND CUTS IN CORPORATE INCOME TAXES[c]
Welfare Costs[a]	$3,356.47	$2,479.64	$1,068.60	$2316.89 (5.4%)	$1319.33 (23.5%)
—per ton of CO_2 reduced	$42.12	$31.30	$13.95	$32.96 (5.3%)	$17.14 (22.9%)

[a] Welfare costs are the negative of the equivalent variation, expressed in billions of 2013 dollars.
[b] Numbers in parentheses in this column are percentage changes from the results in the case with individual income tax cuts alone.
[c] Numbers in parentheses in this column are percentage changes from the results in the case with corporate income tax cuts alone.

applying to the CES. Thus, the nature of recycling of carbon tax revenues is the principal determinant of differences between the distributional impacts of the CES and the carbon tax.

Figure 7.13b contrasts the distributional impacts of an increase in federal gasoline taxes with the equivalently scaled carbon tax.[26] The results are shown for recycling through lump-sum rebates and cuts in individual income tax rates. For the gasoline tax increase, the use- and source-side impacts are both progressive. Hence the overall impact is progressive. The progressivity on the use side reflects low quintile households' smaller share of spending on motor vehicle fuels compared to higher quintiles.[27] As in the case of the carbon tax, source-side impacts of the gasoline tax are highly progressive when revenues are used to finance lump-sum rebates to households, and less progressive when revenues are recycled through cuts in individual income tax rates. As a result, the gasoline tax with individual income tax cuts is less progressive than the gasoline tax with lump-sum rebates.

Figure 7.13b also displays the distributional impacts of the equivalently scaled carbon tax. These impacts largely mirror the impacts from the central case carbon tax, but are much smaller in absolute magnitude,

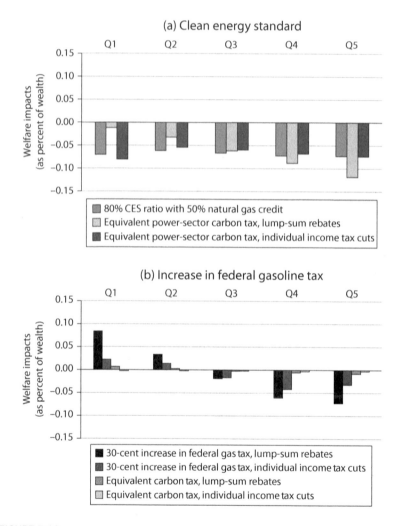

FIGURE 7.13

Welfare Impacts by Quintile under Alternative Climate Policies
(Impacts over the Infinite Horizon)

a reflection of the difference in the level of the carbon tax rate. Paralleling the result for the representative E3 household (examined in chapter 6), for each quintile the welfare costs of the gasoline tax increase are about 10 times greater than the costs of the equivalently scaled carbon tax.

These results reinforce a main theme from earlier chapters: that the method of revenue recycling can be more important to policy outcomes than the choice of main policy type. The nature of revenue recycling can dramatically affect both overall policy costs and their distribution across major economic actors.

■　■　■

The overall attractiveness of climate-change policies depends both on cost-effectiveness and on how the costs are distributed. The latter is relevant to the fairness of the outcome and influences political feasibility.

In this chapter we have concentrated on the distributional dimension—the way policy costs are distributed across industries and household income groups. We find that the distributional pattern depends at least as much on the ways that policy-generated revenues are recycled as on the choice across the main policy types we have investigated.

In the absence of targeted compensation, emissions pricing in the form of a carbon tax or cap-and-trade program tends to give rise to larger adverse costs in the industries that intensively supply or intensively use carbon-based fuels. We find, for example, that under our central case carbon tax, the coal extraction and coal-fired generation industries would experience profit losses of 40 and 62 percent, respectively, in the year 2020, a consequence of significant fuel switching within the power sector. The petroleum refining industry, on the other hand, experiences much smaller profit losses—5 percent in 2020—reflecting the lack of close substitutes for refined products such as gasoline and diesel. The profit losses experienced by carbon-intensive industries apply when carbon revenues are returned through lump-sum rebates and when they are recycled via cuts in marginal tax rates.

We have explored a range of targeted compensation mechanisms to reduce or eliminate the profit losses to the most vulnerable industries. The mechanisms include targeted tax credits under a carbon tax and the free allocation of emissions allowances under a cap-and-trade program. Such compensation could enhance political feasibility by "buying in" industry groups, but it usually entails a sacrifice of cost-effectiveness. A critical issue is the magnitude of this sacrifice. We find that preventing loss of profit in the ten most vulnerable U.S. industries raises costs by about 10 percent relative to the case where all the net revenues from a carbon tax, or all net revenues from auctioned emissions allowances under cap and trade, are recycled through cuts in the individual income tax.

Groups that might otherwise support climate-change policies often express concern about the potential regressivity of CO_2 emissions pricing policies. Our analysis confirms that emissions pricing policies (carbon taxes or cap and trade) do exert a regressive impact in terms of their effects on the prices of goods and services consumed. This stems from the fact that carbon-intensive goods and services (such as electricity, natural gas, and motor vehicle fuels) occupy a larger share of the budgets of low-income households than of more affluent households. However, we find that to gauge the progressivity or regressivity of CO_2 emissions pricing policies, it is critical to consider both use- and source-side impacts. The source-side impacts tend to be progressive, and in several instances they offset all of the regressivity of the use-side effect. We also find that both the scale and the progressivity of the overall (use- plus source-side) impacts depend on the method of recycling.

As with the distribution of impacts across industries, we find that achieving a more even distribution of impacts across households entails a loss of cost-effectiveness. We have examined the extent to which the costs of a carbon tax increase when targeted compensation is introduced in the form of lump-sum payments to the lowest two household quintiles sufficient to prevent a welfare loss to these households. Depending on how remaining net revenues are recycled, such targeted compensation raises the aggregate welfare costs of individual and corporate income tax cuts between 5 and 23 percent.

Alternatives to emissions pricing, clean energy standards, and a gasoline tax exhibit proportional (CES) or progressive impacts (gasoline tax). Under the CES policy, regressive use-side impacts offset progressive source-side impacts, while under the gasoline tax, both the use- and source-side impacts are progressive.

Although we observe some important distributional impacts, it is important to recognize that our analysis does not consider differences across firms within industries, or demographic differences across households within income or expenditure quintiles. Thus, the compensation measures we consider address the average firm in each industry and the average household in each quintile. The extent to which these measures offset adverse impacts to particular firms or individuals can vary significantly.[28]

IV

CONCLUSIONS

———

8

KEY INSIGHTS

Scientific evidence of climate change, and of human activities as a significant cause of such change, continues to mount. And according to a substantial majority of climate scientists, in the absence of expanded global efforts to reduce emissions of greenhouse gases, climate change will become ever more substantial and the associated damages ever more severe.

The U.S. currently lacks a broad and coherent federal policy to bring about reductions in greenhouse gas emissions on a scale consistent with what scientists call for.[1] This book has considered a range of federal-level options that could bring about reductions in energy-related CO_2 emissions: a carbon tax, a cap-and-trade program, a clean energy standard, and an increase in the federal tax on gasoline. These main policy types have gained considerable attention in policy circles. Some of them have been introduced in certain states, but none has been adopted nationwide.

We have applied our numerical model to compare the options, considering their overall impacts as well as the way the impacts are distributed across industries and households. Our approach allows for consistent comparisons across policies by applying a common modeling framework. And our model stands out in its ability to consider the important interactions between these policies and the fiscal system. It recognizes how the costs of climate policies depend on the nature and extent of preexisting taxes and their associated distortions. It also takes account of how recycling of policy-generated revenues can reduce costs, particularly when the revenues are used to cut prior taxes.

MAJOR FINDINGS

We have examined the impacts of the climate policies along several dimensions, assessing the environmental benefits (avoided climate damages and other, nonclimate benefits) and the economic costs. We have estimated the costs to the economy as a whole and the distribution of these costs across industries and household income groups. Climate policy has the potential to impose disproportionate impacts on particular industries or household groups. This has motivated our attention to policy designs that avoid or reduce uneven impacts on profits or household welfare. By evaluating the additional economic sacrifice associated with such policies, we have been able to assess the trade-offs between cost-effectiveness and the evenness of distributional outcomes. Here we summarize the results along each of these dimensions.

Net Benefits

Each of the four main policy types has the potential to generate benefits in excess of the economic costs. The sign and scale of the net benefits depend on several key factors. One is the assumed value of the social cost of carbon (SCC)—the climate-related damage from an additional ton of emissions of CO_2 (or equivalently, the benefit from an avoided ton of CO_2 emissions).[2] A second is the policy's stringency—e.g., the size of the tax involved or the tightness of the emissions cap. A third is the nature of revenue recycling.

We have evaluated the net benefits for a family of carbon tax policies involving a tax rate that increases at four percent per year after a 3-year phase-in. As indicated in chapter 5, the carbon tax yields climate benefits in excess of the economic costs across a wide range of initial tax levels and estimates for the SCC. Only under very low estimates for climate damages do we find carbon taxes with negative net benefits. For central and high estimates of the SCC, we find that carbon taxes yield net benefits even for the highest carbon tax profile considered, which yields cumulative emissions reductions of 45 percent.[3]

In addition to reducing CO_2 emissions, a carbon tax yields reductions in emissions of local air pollutants and thereby yields nonclimate benefits in the form of avoided damages to health. From our estimates, the nonclimate benefits are considerably larger than the climate benefits. Thus, accounting for the nonclimate benefits significantly raises the carbon tax's estimated net benefits. For all of the carbon tax policies that we have considered—even the most stringent (and costly)—the nonclimate benefits alone exceed the policy costs.

Equivalently scaled economywide cap-and-trade systems yield net benefits under the same conditions. This reflects the very similar economic incentives generated by carbon taxes and cap and trade.

CES policies also produce significant net benefits (exclusive of nonclimate benefits) over the ranges considered in this text. In contrast, the range of gasoline tax increases considered do not produce climate benefits in excess of the policy costs, no matter what form of revenue recycling is involved. However, once one accounts for the nonclimate benefits, the gasoline tax policies yield net benefits. This is the case under all forms of revenue recycling considered. The nonclimate benefits from higher gasoline taxes are mainly in the form of avoided health costs from reduced local air pollution. These benefits are considerably larger than the policy's climate benefits.

Costs

We find that the gross costs[4] of policies depend crucially on the stringency of the policy, that is, the extent of the emissions-reduction target. Stringency can determine whether, after recycling the revenues, the policy costs are positive. As discussed in previous chapters, there has been much interest in whether a revenue-neutral carbon tax might yield a double dividend: both a climate benefit (avoided climate damages) and a reduction in the costs of the tax system. Chapter 5's numerical assessments obtained the double dividend for a carbon tax when two conditions were jointly met: the revenues were recycled through cuts in corporate taxes, and the level of stringency was very low. The result meshes with the principle, outlined in the theoretical framework of chapter 2, that the double dividend requires that the revenue-neutral policy reduce a preexisting nonenvironmental inefficiency in the tax system. In the numerical assessment in chapter 5, the carbon tax combined with recycling via cuts in the corporate income tax helped reduce the excessive taxation (on efficiency grounds) of capital relative to labor.

Under parameter values for our central case, the gross costs of the revenue-neutral policy with corporate income tax cuts were negative for carbon tax profiles that yielded cumulative emissions reductions of 10 percent or less. For more aggressive carbon tax time profiles, gross costs became positive: the relative narrowness of the carbon tax compared to the income tax it replaces ultimately was the dominant cost factor. Surely the possibility of a "free" (zero or negative gross cost) carbon tax policy offers political attractions. However, reducing the policy stringency enough to obtain the double dividend involves a sacrifice of the policy's net benefits, since it implies smaller environmental gains.

Stringency not only affects a policy's absolute costs but also can alter its costs relative to that of other policies. As indicated in chapter 6, when the emissions-reduction target is not very stringent, the CES achieves those targets at lower cost than an equivalent power-sector-only carbon tax— even when the carbon tax exploits the revenue-recycling effect through cuts in individual income tax rates. Economists have tended to view carbon taxes and other emissions pricing policies as more cost-effective than intensity standards, of which the CES is an example. However, this viewpoint requires modification once one accounts for tax interactions. In chapter 6 we compared the costs of a CES with that of an equivalently scaled power-sector-only carbon tax. While the CES is at a disadvantage relative to the carbon tax in terms of its ability to encourage conservation (demand reduction), it has the advantage of producing a smaller (adverse) tax-interaction effect. For quite modest reduction targets (reductions below 15 percent), this advantage outweighs the CES's conservation-related disadvantage. But for policies of higher stringency (15–21 percent reductions), emissions pricing policies are more cost-effective provided that they involve revenue recycling in the form of individual income tax cuts. Under policies of even greater stringency (reductions above 21 percent), the CES's conservation-related disadvantage gains greater weight, and the CES is more costly than equivalently scaled power-sector-only emissions pricing even when the emissions pricing policies involve less cost-effective forms of recycling.

The gasoline tax spans much less of the economy's sources of CO_2 emissions than the other policies. Emissions from household use of motor vehicle fuels account for roughly 9–10 percent of total U.S. carbon emissions in the years 2016–2050. Thus, even fairly large increases in the gasoline tax produce smaller reductions in CO_2 emissions than those considered for the other policies. For example, a 60-cent increase in the federal gasoline tax reduces emissions by less than 1.5 percent.[5] Because of its narrower base, the costs per ton of CO_2 emissions abatement are considerably higher than under the other three main policies considered, regardless of the level of stringency or the form of revenue recycling. The higher costs per ton make it difficult to justify a gasoline tax increase as a climate policy *per se*. On the other hand, as mentioned earlier and examined more closely in chapter 6, a higher gasoline tax would yield substantial health benefits associated with the policy-generated reductions in local air pollutants—benefits large enough to outweigh the policy's costs.

The significance of stringency for policy rankings also emerged in chapter 5's comparison of a narrow, power-sector-only cap-and-trade policy with a broader, economywide one. The discussion of the gasoline tax earlier suggests that a narrower base generally works against cost-effectiveness. However, electricity-only emissions pricing policies sometimes provide an exception to this general rule. As with the CES policy, considerations of fiscal interactions prove to be important. Our analysis shows that because electricity demands are relatively elastic, the electric power sector tends to generate a smaller tax-interaction effect than other sectors of the economy. As a consequence, under the broader cap-and-trade policy, the share of the economy's emissions reductions achieved by the electric power sector is too low from an efficiency point of view. When the emissions reduction target is modest (requiring reductions of about 20 percent), the narrower cap-and-trade policy comes closer to the optimum, and this is reflected in its lower cost. It is important to recognize, however, that for policies involving greater reductions, the broader policy is more cost-effective, sometimes considerably so.[6]

Distribution of Impacts

Clearly, the fairness of policies depends on how the impacts of policies are distributed. And the distribution of costs or benefits also affects political feasibility: highly mobilized groups that are particularly harmed by a given policy can have enough influence to prevent that policy from becoming law.

Industry Impacts

Results from our policy simulations reveal that, in the absence of specific compensation components, the climate policies yield costs that are highly concentrated among a relatively few carbon-intensive industries. We identified 10 industries as most vulnerable: coal-fired electricity generation, coal mining, natural gas extraction, other fossil electricity generation, natural gas distribution, electric transmission and distribution, pipeline transportation, mining support activities, petroleum refining, and railroad transportation. Under our central case economywide carbon tax, for these industries the losses in profit in the year 2020 range from 4 to 75 percent. The profit losses in other industries, however, are considerably smaller, with average losses of about 1 percent. Under the CES and the electric power-sector-only cap-and-trade policy, the profit losses are concentrated only in electricity-related industries; hence the overall profit losses are smaller, reflecting the narrow coverage of each policy. The federal gasoline tax has a small profit impact on the petroleum refining industry (about 2 percent), but only imperceptible impacts on profits of other industries. This reflects the narrow base of the gasoline tax policy we consider.[7]

In chapter 7 we investigated a number of ways that climate policies could be designed to reduce the unevenness of the percentage losses of profit or to prevent profit losses in the most vulnerable industries. Within a cap-and-trade system, the free allocation of emissions allowances to firms in certain industries can reduce or eliminate the adverse profit impact. Within a carbon tax policy, there is a similar (but less recognized) way to provide compensation: offer tradable exemptions to the carbon tax, to firms in the vulnerable industries, for their emissions up to some specified level. As with free allowances under cap and trade, the tradable exemptions within a carbon tax policy could differ across industries or even across firms within an industry. Alternatively, the federal government could reduce the corporate tax rate in industries that would be particularly impacted by emissions pricing.

As indicated in chapter 7, compensation tends to reduce cost-effectiveness. Free allowances and tax exemptions reduce the gross revenue that the government can bring in from these emissions pricing policies and can thereby reduce the potential to exploit the revenue-recycling effect. For example, compared with the no-compensation case, the welfare costs of a carbon tax with individual income tax recycling are 10 percent higher when tax exemptions are used to avoid profit losses in the ten most vulnerable industries. The extent to which this increases fairness (or political feasibility), and whether this is a reasonable price to pay this increase, is a subjective matter. From our perspective, the cost increase seems minor.

Household Impacts

Ultimately, all the costs of climate-change policies are borne by people. Thus, to assess policy fairness, it is especially important to consider the impacts on individuals. We have concentrated on how the policy costs are distributed across quintiles of households, where the households are ranked by income or expenditure.

Our modeling framework enables us to decompose the impacts on each household group (or quintile) into use-side and source-side impacts, where the former is the welfare impact stemming from changes in the prices of goods or leisure consumed and the latter is the impact from changes in wage and capital income and nominal government transfers. Our approach also reveals how the distribution of these impacts changes over time.

A carbon tax increases the relative prices of goods and services that occupy a larger share of spending by individuals in the lower income or expenditure quintiles. As a result, the use-side impact is regressive. The particular form of revenue recycling can alter the extent of regressivity, but the use-side impacts remain regressive under all the revenue-recycling options considered.

In contrast, the source-side impacts are generally progressive. Recycling through lump-sum rebates of equal amounts to each household yields a very progressive source-side impact: the given rebate is larger relative to income for the low-income households, and other source-side impacts of the carbon tax (e.g., impacts on after-tax returns to wage and capital income) do not offset this progressive impact. Recycling through cuts in corporate income tax rates produces a less progressive impact because it favors capital income, which represents a disproportionally large share of the income of the more affluent households.

In most recycling cases, the source-side impact dominates the use-side impact, so that the overall impact on welfare is progressive. The extent of progressivity of the overall impact depends on the recycling method; this largely reflects the influence of recycling on the progressivity of the source-side impact.

Our results suggest trade-offs between cost-effectiveness and fairness (where fairness is measured by the evenness of policy outcomes across household groups). The recycling methods that lead to the most cost-effective outcomes (cuts in individual and corporate income tax rates) also are the most regressive overall.[8] We have evaluated the trade-offs by measuring the efficiency sacrifice associated with reserving some revenue for recycling in the form of lump-sum compensation to low-income households, as opposed to recycling all the revenue through cuts in individual or corporate income tax rates. We find that the costs of preventing losses to low-income quintiles, on average, are generally low: costs rise by 5 and 23 percent relative to the pure individual or corporate income tax rate cut policies, respectively.

STRENGTHS AND LIMITATIONS OF THIS ANALYSIS

Many of the conclusions suggest the importance of interactions between climate policy and the fiscal system. We have emphasized the ability of our model to account for these interactions. But the model has other features that are critical to its evaluation of the policy options that we have considered. One important feature is its attention to the costs of installing or removing physical capital. As indicated in the introduction, recognizing these costs is critical to evaluating the profit impacts of policies in particular industries. Indeed, because most economywide models of climate policy ignore these costs, they do not attempt to indicate the impacts on profits.

A second key feature is the ability to consider specific design features of climate policies, including various revenue recycling options, alternative specifications for the range of industries covered, potential industry-specific tax deductions or tax rate cuts, potential credits for natural gas under the CES, and various forms of compensation to particular household income groups.

A third contribution, provided through our disaggregated household model, is the ability to decompose the household welfare impacts into source-side and use-side impacts as well as to display how these impacts change over time.

We have examined the climate policies under a wide range of assumptions about initial economic conditions and the parameters governing producer and consumer behavior. Our results remain highly robust to the alternative assumptions, and this lends confidence to our findings.

Still, our framework has its limitations, and it is worth considering the extent to which these limitations could affect our findings. We present four limitations here. The first two might bias our model to overstate the benefits from the climate policies we consider, while the last two suggest a bias toward understatement.

First, the E3 model assumes that firms are purely competitive; it abstracts from the imperfect competition that can be important in some industries. Although there are some excellent partial equilibrium studies of climate-change policy that address imperfect competition in a particular industry[9], to our knowledge there is at present no multisector general equilibrium analysis of U.S. climate-change policy that has incorporated imperfect competition. Imperfect competition could imply somewhat smaller net benefits from any given climate policy: to the extent that output in imperfectly competitive industries is already being restricted because of imperfect competition, the efficiency gain from additional conservation caused by climate policy would be lowered. The degree to which adding imperfect competition would change our results remains an open question, but we would not expect this effect to be especially large given that pure competition closely approximates market structure in the industries such as services, trade, and construction that account for over 50 percent of gross output.

A second limitation is that our model lacks the detail on production technologies of various "bottom-up" models.[10] In particular, our electric power sector does not distinguish between different types of generators within our generating industries. Specifically, the E3 model does not distinguish between turbine or combined-cycle natural gas generation; nor does it distinguish different renewable energy sources, such as solar-powered, wind-generated, and geothermal-powered electricity. Further, it does not have details on departures from marginal cost pricing in some industries; for example, the use of two-part tariffs and average cost pricing in U.S. natural gas markets or average cost pricing in some electricity markets in the country.[11] Our model's inattention to average cost pricing could imply an upward bias in our estimates of the efficiency gains from emissions pricing to the extent that average cost pricing already leads to prices above marginal cost. Thus, accounting for average cost pricing could lower somewhat our estimates of the net benefits from the various climate policies, though we would not expect this to have a large quantitative impact.

Third, our model treats the rate of technological change as exogenous. In reality, climate policies affect incentives to engage in research toward the invention of new technologies, and this can affect the rate of technological progress in energy-related and other industries. Although several studies have explored how climate policies might influence the rate of technological change,[12] the quantitative impact remains highly uncertain. Still, it is reasonable to conclude that accounting for policy-induced technological change would lower the costs of meeting the various CO_2 abatement targets considered in this book.

Finally, in assessing the federal-level policies, we have not accounted for the fact that the broad-based policies that we have focused on would make some existing climate policies redundant. A broad-based carbon tax, for example, could make unnecessary several current climate-related policies, many of which are widely viewed as inflexible and inefficient.[13] If these inefficient policies were removed by Congress as part of a bill for a broad-based carbon tax, our model would underestimate the positive benefits of the policy because it does not take into account the benefits from eliminating these policies. This limitation suggests that our positive results are conservative.

It is difficult to discern the overall impact of these limitations. However, we would not expect the quantitative importance of any one of these limitations to be exceptionally large. Moreover, particular limitations work in opposite directions and thus are offsetting. Based on these considerations, we believe that, despite these limitations, the central findings that we have outlined earlier in this chapter remain valid.

FINAL WORDS

Our analysis reveals that appropriately designed climate policies can produce significant reductions in CO_2 emissions and yield climate-related benefits that exceed the gross costs. The benefits include avoided future climate-related damages as well as benefits associated with avoided air pollution. Emissions pricing—either in the form of a carbon tax or a cap-and-trade program—with judicious use of revenues tends to be the most cost-effective approach among those considered in this book. However, whether the cost-effectiveness advantage is realized depends on design elements, including policy stringency and the method of revenue recycling.

Each of the four main policy types produces uneven profit impacts across industries, regardless of the recycling option, with the largest adverse impacts (in percentage terms) applying to the coal mining industry and to coal-fired electricity generators. The impacts across household quintiles reflect a combination of use-side impacts (arising from changes in the prices of goods and services consumed) and source-side impacts (stemming from changes in the returns to factors of production). While the use-side impacts are generally regressive, the source-side impacts are progressive. The overall impact is generally either progressive or proportional, depending on the recycling option. In general, the adverse profit impacts on the most vulnerable industries and the negative impacts on low-income household groups can be eliminated at relatively low cost to the overall economy.

We hope that these findings will lead to more informed discussions by policy makers and advance the efforts to produce a more environmentally effective, efficient, and fair approach to climate-change policy in the U.S.

APPENDIX A

FISCAL INTERACTIONS AND THE COSTS OF ENVIRONMENTAL POLICIES

Here we present a simple analytical framework for assessing the impacts of environmental policies in economies with preexisting distortionary taxes. The framework borrows from the analytical general equilibrium analysis in Goulder, Parry, and Burtraw (1997), which we hereafter refer to as GPB.

THE MODEL

A representative agent model is assumed in which household utility is

$$U(X_1, X_2, \ell) + V(Z) \qquad \text{(A.1)}$$

where $U(\bullet)$ is quasi-concave, $V(\bullet)$ is concave, and both functions are continuous. Here X_1 and X_2 are market goods, ℓ is leisure or nonmarket time, and Z is the quality of the environment. Separability between environmental and nonenvironmental goods implies that the demand functions for X_1, X_2, and ℓ are independent of Z.

Both X_1 and X_2 are produced by competitive firms using labor as the only input. The marginal product of labor in both industries is constant and unaffected by environmental quality. Normalizing units to imply transformation rates of unity, we can write the economy's resource constraint as

$$\bar{l} = X_1 + X_2 + \ell \qquad \text{(A.2)}$$

where $\bar{l}$ is the household time endowment ($\bar{l} - \ell$ is the labor supply).

The production of X_1 causes waste emissions that harm the environment; that is,

$$Z = Z(X_1) \tag{A.3}$$

where $Z_{X_1} < 0$. From (A.1) and (A.3), we can define marginal environmental damages from production of X_1 in terms of dollars by

$$D(X_1) = \frac{1}{\lambda} V' Z_{X_1} \tag{A.4}$$

where λ is the marginal utility of income. Because V is concave ($V' < 0$), $D'(X_1) \geq 0$. It is assumed that in the absence of policy intervention, there is no internalization of environmental damages by firms or households.

Finally, the government has an exogenous total revenue requirement G^T, levies a proportional tax of τ_L on labor income, and regulates X_1. For our purposes, we assume that government revenue is returned to households as a lump-sum transfer.

IMPACTS OF POLLUTION TAXES, WHEN RECYCLING IS VIA CUTS IN MARGINAL TAX RATES

Consider a revenue-neutral policy involving a new environmental tax τ_{X_1} per unit of output X_1. The environmental tax is accompanied by a reduction in the labor tax τ_L, where the reduction is such as to make the overall policy revenue-neutral. Normalizing the gross wage to unity, we can express the household budget constraint as

$$\left(1 + \tau_{X_1}\right) X_1 + X_2 = (1 - \tau_L)(\overline{l} - \ell) + G^T \tag{A.5}$$

Households are assumed to maximize utility (A.1) subject to their budget constraint (A.5), taking environmental quality as given. This yields the first-order conditions

$$U_{X_1} = \left(1 + \tau_{X_1}\right)\lambda \qquad U_{X_2} = \lambda \qquad U_\ell = (1 - \tau_L)\lambda \tag{A.6}$$

From (A.5) and (A.6) we can implicitly derive the (uncompensated) demand functions

$$X_1\left(\tau_{X_1},\tau_L\right) \quad X_2\left(\tau_{X_1},\tau_L\right) \quad \ell\left(\tau_{X_1},\tau_L\right) \tag{A.7}$$

Government revenues are the sum of labor and pollution tax revenues. Therefore, government budget balance requires

$$G^T = \tau_L\left(\overline{l}-\ell\right)+\tau_{X_1}X_1 \tag{A.8}$$

Totally differentiating the resource constraint (A.2) yields the following condition for the aggregate quantity effects of the policy change:

$$\frac{d\ell}{d\tau_{X_1}}+\frac{dX_1}{d\tau_{X_1}}+\frac{dX_2}{d\tau_{X_1}}=0 \tag{A.9}$$

where

$$\frac{d\ell}{d\tau_{X_1}}=\frac{\partial\ell}{\partial\tau_{X_1}}+\frac{\partial\ell}{\partial\tau_L}\frac{d\tau_L}{d\tau_{X_1}} \tag{A.10}$$

and where analogous expressions apply for $dX_1/d\tau_{X_1}$ and $dX_2/d\tau_{X_1}$.

Differentiating (A.1) and using (A.3) and (A.7), we obtain the following expression for the policy change's impact on utility:

$$\frac{dU}{d\tau_{X_1}}=\left(U_{X_1}+V'Z_{X_1}\right)\frac{dX_1}{d\tau_{X_1}}+U_{X_2}\frac{dX_1}{d\tau_{X_1}}+U_L\frac{d\ell}{d\tau_{X_1}}$$

Substituting the expressions for marginal environmental damage (A.4), marginal utility from consumption (A.6), and the aggregation property (A.9) gives the following expression for the policy's money-equivalent impact:

$$\frac{1}{\lambda}\frac{dU}{d\tau_{X_1}}=-\left(D-\tau_{X_1}\right)\frac{dX_1}{d\tau_{X_1}}-\tau_l\frac{d\ell}{d\tau_{X_1}} \tag{A.11}$$

We now decompose the second term on the right-hand side of (A.11). Totally differentiating the government budget constraint (A.8), using (A.7) and (A.10), we can obtain

$$\frac{d\tau_L}{d\tau_{X_1}} = -\frac{X_1 + \tau_{X_1} dX_1/d\tau_{X_1} - \tau_L \partial\ell/\partial\tau_{X_1}}{\overline{1} - \ell - \tau_L \partial\ell/\partial\tau_L} \tag{A.12}$$

This expression is the reduction in labor tax that can be financed by a marginal increase in the environmental tax, while maintaining budget balance. Substituting (A.10) and (A.12) in (A.11) gives

$$\frac{1}{\lambda}\frac{dU}{d\tau_{X_1}} = \underbrace{(D - \tau_{X_1})\left(-\frac{dX_1}{d\tau_{X_1}}\right)}_{d\mathcal{W}^P} + \underbrace{\text{MEB}\left(X_1 + \tau_{X_1}\frac{dX_1}{d\tau_{X_1}}\right)}_{d\mathcal{W}^R} - \underbrace{(1 + \text{MEB})\tau_L\frac{\partial\ell}{\partial\tau_{X_1}}}_{\partial\mathcal{W}^I} \tag{A.13}$$

where

$$\text{MEB} = \frac{\tau_L \partial\ell/\partial\tau_L}{\overline{1} - \ell - \tau_L \partial\ell/\partial\tau_L} \tag{A.14}$$

The numerator in (A.14) is the partial equilibrium welfare loss from a marginal increase in the labor tax. This is the increase in leisure multiplied by τ_L, the wedge between the gross and net wages. The denominator is the partial equilibrium increase in government revenues from a marginal increase in the labor tax (from differentiating $\tau_L(\overline{1} - \ell)$ with respect to τ_L). Therefore, MEB represents the marginal excess burden of labor taxation, the (partial equilibrium) efficiency cost of raising an additional dollar of revenue by increasing the labor tax.

Equation (A.13) decomposes the welfare impact of the policy change into three components. The first is the (marginal) *Pigouvian gain* $d\mathcal{W}^P$. This gain represents the difference, at the margin, between the environmental benefit and primary cost. The marginal environmental benefit is $D(-dX_1/d\tau_{X_1})$ and marginal *primary cost* is $\tau_{X_1}(-dX_1/d\tau_{X_1})$.

The second and third terms in (A.13) refer to fiscal interactions. The second is the gain from the (marginal) *revenue-recycling effect* $d\mathcal{W}^R$. This is the product of the efficiency value per dollar of tax revenue (the marginal excess burden of taxation) and the incremental pollution tax revenue. The third term is the (marginal) *tax-interaction effect* $\partial\mathcal{W}^I$. When X_1 and leisure are substitutes, an increase in the demand price of X_1 increases leisure, which exacerbates the welfare cost of the labor tax by $\tau_L \partial\ell/\partial\tau_L$.

This also reduces labor tax revenues by $\tau_L \partial \ell / \partial \tau_L$. The tax-interaction effect is the welfare loss from these two impacts.

We now compare the revenue-recycling and tax-interaction effects. Using (A.13) and (A.14), we can express $\partial \mathcal{W}^I$ as $\mathrm{MEB}(\bar{l} - \ell)(\partial \ell / \partial \tau_{X_1}) / (\partial \ell / \partial \tau_L)$. Substituting the Slutsky equations and making use of the Slutsky symmetry property, we can obtain

$$\partial \mathcal{W}^I = \frac{-\mathrm{MEB}(\bar{l} - \ell)\left(\dfrac{\partial X_1^c}{\partial \tau_L} + \dfrac{\partial \ell}{\partial I} X_1\right)}{\dfrac{\partial \ell^c}{\partial \tau_L} - \dfrac{\partial \ell}{\partial I}(\bar{l} - \ell)} \tag{A.15}$$

where c denotes a compensated coefficient and I is disposable household income. Differentiating (A.2) yields

$$\frac{\partial \ell^c}{\partial \tau_L} = -\left\{\frac{\partial X_1^c}{\partial \tau_L} + \frac{\partial X_2^c}{\partial \tau_L}\right\} \tag{A.16}$$

Substituting (A.16) in (A.15) and using (A.2), we obtain equation (A.17), which gives the following approximation for the tax-interaction effect:

$$\partial \mathcal{W}^I = \phi_{X_1} \mathrm{MEB} \cdot X_1 \qquad \phi_X = \frac{\eta_{X_1 \ell}^c + \eta_U}{\dfrac{X_1}{X_1 + X_2} \eta_{X_1 \ell}^c + \dfrac{X_2}{X_1 + X_2} \eta_{X_2 \ell}^c + \eta_U} \tag{A.17}$$

where

$$\eta_{X_1 \ell}^c = -\frac{\partial X_1^c}{\partial \tau_L} \frac{1 - \tau_L}{X_1} \quad \eta_{X_2 \ell}^c = -\frac{\partial X_2^c}{\partial \tau_L} \frac{1 - \tau_L}{X_2} \quad \text{and} \quad \eta_U = -\frac{\partial \ell}{\partial I} \frac{(1 - \tau_L)(\bar{l} - \ell)}{(\bar{l} - \ell)} \tag{A.18}$$

and where $\eta_{X_1 \ell}^c$ and $\eta_{X_2 \ell}^c$ are the compensated elasticities of demand for X_1 and X_2 with respect to the price of leisure and η_U is the income elasticity of labor supply. Here ϕ_X is a measure of the degree of substitution between X_1 and leisure relative to that between aggregate consumption and leisure. ϕ_X equals unity when X_1 and X_2 are equal substitutes for leisure ($\eta_{X_1 \ell}^c = \eta_{X_2 \ell}^c$) and is greater (less) than unity when X_1 is a relatively strong (weak) substitute for leisure, that is, when $\eta_{X_1 \ell}^c$ is greater (less) than $\eta_{X_2 \ell}^c$. When $\eta_{X_1 \ell}^c = \eta_{X_2 \ell}^c$, equation (A.17) reduces to $\partial \mathcal{W}^I = \mathrm{MEB} \cdot X_1$.

Equation (A.13) indicates that under these circumstances the tax-interaction and revenue-recycling effects exactly cancel each other if pollution abatement is incremental, that is, if $\tau_{X_1} = 0$. However, for more than incremental abatement (i.e., for $\tau_{X_1} > 0$), the tax-interaction effect is larger than the revenue-recycling effect (by the amount $\mathrm{MEB}\tau_{X_1}\, dX_1/d\tau_{X_1}$). The intuition is as follows. The pollution tax affects the relative prices of consumer goods, "distorting" the household's consumption choice as well as its labor-leisure choice. Recycling the revenues helps return the real wage to its original value and thereby mitigates the labor market distortion, but such recycling does not undo the change in relative consumer good prices and the associated "distortion" in consumption. For this reason the revenue-recycling effect only partly offsets the tax-interaction effect when the level of abatement is nonincremental.

Thus, if the environmentally damaging good is an average substitute for leisure, preexisting taxes cause the efficiency gains from environmental taxes to be lower than in a first-best setting with no distortionary taxes. However, if the environmentally damaging good is a sufficiently weak substitute for leisure (i.e., unless ϕ_X is sufficiently below unity), the opposite result would apply.

Equation (A.13) also implies that when the environmentally damaging good is an average substitute for leisure, the gross costs of the environmental tax are positive: the double dividend is not obtained. Gross cost is primary cost plus the cost from the tax-interaction effect minus the benefit from the revenue-recycling effect. In general, gross cost is positive. And when the environmentally damaging good is an average leisure substitute, the beneficial revenue-recycling effect is smaller than the adverse tax-interaction effect. Hence gross costs are positive.

IMPACTS IN THE ABSENCE OF THE REVENUE-RECYCLING EFFECT

Now consider the case where the revenues from the pollution tax are returned in lump-sum fashion rather than in the form of marginal rate cuts. This case is formally the same as the case of cap and trade with freely allocated emissions permits.

Let the pollution tax be represented as τ_{X_1}. Let the revenue from this tax be expressed by $\pi = \tau_{X_1} X_1$. One can also regard π as the rent from an equivalent cap-and-trade system in which allowances are given out free. Under both interpretations, π becomes an exogenous lump-sum component of income in the household budget constraint. The household demand functions can now be summarized by

$$X_1\left(\tau_{X_1},\tau_L,\pi\right) \qquad Y\left(\tau_{X_2},\tau_L,\pi\right) \qquad \ell\left(\tau_{X_1},\tau_L,\pi\right) \qquad \text{(A.7a)}$$

The key difference from the marginal tax rate recycling case earlier is that the policy does not produce a revenue-recycling effect. In both the lump-sum rebate tax policy and the cap-and-trade system with free allocation, the government budget constraint is[1]

$$G^T = \tau_L(\bar{l} - \ell) \qquad \text{(A.8a)}$$

Again we consider a revenue-neutral incremental increase in τ_{X_1}. We can obtain analogous expressions for the aggregation property (A.9) and welfare change expression (A.11), except that expression (A.10) becomes

$$\frac{d\ell}{d\tau_{X_1}} = \frac{\partial \ell^I}{\partial \tau_{X_1}} + \frac{\partial \ell}{\partial \tau_L}\frac{d\tau_L}{d\tau_{X_1}} \qquad \text{(A.10a)}$$

where $\dfrac{\partial \ell^I}{\partial \tau_{X_1}} = \dfrac{d\ell}{d\tau_{X_1}} + \dfrac{\partial \ell}{\partial \pi}\dfrac{d\pi}{d\tau_{X_1}}$ and analogous expressions apply for $dX_1/d\tau_{X_1}$

and $dX_2/d\tau_{X_1}$ from equation (A.9). Here I denotes a coefficient that takes into account the income effect from the lump-sum revenue or rent to the household. Differentiating the government budget constraint (A.8a) using (A.7a) and (A.10a) yields

$$\frac{d\tau_L}{d\tau_{X_1}} = -\frac{\tau_L \partial \ell^I/\partial \tau_{X_1}}{\bar{l} - \ell - \tau_L \partial \ell/\partial \tau_L} \qquad \text{(A.12a)}$$

Since $\partial \ell^I/\partial \tau_{X_1}$ is (in general) positive, revenue neutrality requires an increase, rather than a decrease, in the labor tax.

Applying a procedure analogous to the one used to derive equation (A.13) earlier, we arrive at the following expression for the general equilibrium welfare change from the policy:

$$\frac{1}{\lambda}\frac{dU}{d\tau_{X_1}} = \underbrace{\left(D-\tau_{X_1}\right)\left(-\frac{dX_1}{d\tau_{X_1}}\right)}_{\partial \mathcal{W}^P} - \underbrace{(1+\text{MEB})\tau_L\frac{\partial \ell}{\partial \tau_{X_1}}}_{\partial \mathcal{W}^I}$$

(A.13a)

The difference between (A.13a) and (A.13) is the absence of the revenue-recycling effect. Under an environmental tax with lump-sum recycling (or a cap-and-trade system in which allowances are not auctioned),[2] there is no revenue-recycling effect to counteract the tax-interaction effect.

WELFARE IMPLICATIONS OF POLICY CHOICE

The presence or absence of the revenue-recycling effect affects the efficiency outcomes of environmental policies in important ways. The presence or absence of the revenue-recycling effect affects the relative overall efficiency gains of policies: for comparably scaled environmental tax or cap-and-trade policies, the efficiency gains are larger when the revenue-recycling effect is exploited.

In addition, the sign of the efficiency impact can depend on whether the revenue-recycling effect is present. An efficiency improvement will occur if and only if the combination of the Pigouvian gain (the environmental benefit net of primary cost) and the revenue-recycling effect (if applicable) is larger than the tax-interaction effect. The revenue-recycling effect may be necessary to meet this condition.

Note that the tax-interaction effect is nonincremental, even at the first incremental amount of abatement, as can be seen from equation (A.13) or (A.13a). This means that when the revenue-recycling effect is absent, the Pigouvian gain must exceed a certain positive value to allow an efficiency improvement.

E3 MODEL STRUCTURE IN DETAIL

This appendix offers a complete and detailed description of the structure of the E3 model, complementing the overview offered in chapter 3.

PRODUCER BEHAVIOR

A representative firm in each industry produces distinct gross output X, using capital K, labor L, energy intermediate inputs $E_1,...,E_{N_e}$, and nonenergy (or materials) intermediate inputs $M_1,...,M_{N_m}$.[1] Producers choose variable inputs to minimize costs and make investment decisions to maximize the value of the firm—the present value of after-tax dividends minus new share issues.

Production

Output from each industry stems from a nested structure of constant-elasticity-of-substitution production functions. Figure 3.2 displays this structure. At each node, an aggregation function creates a composite of the inputs immediately below that node. At the lowest part of the nest, each of the inputs $E_1,...,E_{N_e}$ and $M_1,...,M_{N_m}$ is a composite of the good produced by the domestic industry and the good produced by its foreign counterpart,

$$E_j = \gamma_{e_j}^{\frac{1}{\rho_{e_j}}} \left[\alpha_{e_j} (E_j^d)^{\rho_{e_j}} + (1-\alpha_{e_j})(E_j^f)^{\rho_{e_j}} \right]^{\frac{1}{\rho_{e_j}}} \tag{B.1}$$

$$M_j = \gamma_{m_j}^{-\frac{1}{\rho_{m_j}}} \left[\alpha_{m_j} (M_j^d)^{\rho_{m_j}} + (1-\alpha_{m_j})(M_j^f)^{\rho_{m_j}} \right]^{\frac{1}{\rho_{m_j}}} \tag{B.2}$$

where the superscripts d and f refer to domestically produced and foreign-made inputs.

At the next level, intermediate inputs of energy and of materials are aggregated into an energy composite E and a materials composite M.

$$E = \gamma_{\bar{e}}^{-\frac{1}{\rho_{\bar{e}}}} \left[\sum_{j=1}^{N_e} \alpha_{\bar{e},j} E_j^{\rho_{\bar{e}}} \right]^{\frac{1}{\rho_{\bar{e}}}} \tag{B.3}$$

$$M = \gamma_{\bar{m}}^{-\frac{1}{\rho_{\bar{m}}}} \left[\sum_{j=1}^{N_m} \alpha_{\bar{m},j} E_j^{\rho_{\bar{m}}} \right]^{\frac{1}{\rho_{\bar{m}}}} \tag{B.4}$$

At the next level of the nest, the energy and material composites are aggregated into an intermediate input composite using function h:

$$h(E,M) = \gamma_h^{-\frac{1}{\rho_h}} \left[\alpha_h E^{\rho_h} + (1-\alpha_h)M^{\rho_h} \right]^{\frac{1}{\rho_h}} \tag{B.5}$$

One level higher, the intermediate input composite $h(E, M)$ is aggregated with labor using function g:

$$g(L,h(E,M)) = \gamma_g^{-\frac{1}{\rho_g}} \left[\alpha_g L^{\rho_g} + (1-\alpha_g)h(E,M)^{\rho_g} \right]^{\frac{1}{\rho_g}} \tag{B.6}$$

The capital input K is a composite of two types of capital, structures K^s and equipment/intellectual property products K^e:

$$K = k(K^s, K^e) = \gamma_k (K^s)^{\alpha_k} (K^e)^{1-\alpha_k} \tag{B.7}$$

For brevity, we will often refer to K^e simply as equipment, though in fact it captures inputs of intellectual property products as well.

At the top of the production nest, the composite capital input is combined with the labor-intermediate input composite to create gross output X:

$$X = f(K, g(L,h(E,M))) = \gamma_f^{-\frac{1}{\rho_f}} \left[\alpha_f K^{\rho_f} + (1-\alpha_f)g(L,h(E,M))^{\rho_f} \right]^{\frac{1}{\rho_f}} \tag{B.8}$$

At each level of the nest, the ρ parameters are simple translations of the elasticity of substitution $\sigma : \rho = (\sigma - 1)/\sigma$. Given the ρ parameters, the α and γ parameters are calibrated to be consistent with benchmark data inputs for $E_1^d, ..., E_{N_e}^d, E_1^f, ..., E_{N_e}^f, M_1^d, ..., M_{N_m}^d, M_1^f, ..., M_{N_m}^f, L, K^s$, and K^e for each industry. Data sources and primary parameters (such as σ) are described in detail in appendix C. With the exception of the crude oil extraction industry (described later), each industry exhibits constant returns to scale at each level of the nested production function.

Capital adjustment costs are modeled as the sacrifice of output associated with the process of investing in structures and equipment. Specifically, net output Y is equal to gross output minus the adjustment costs

$$Y = X - \phi(I^s/K^s) \bullet I^s - \phi(I^e, K^e) \bullet I^e \tag{B.9}$$

where $\phi(I/K) \bullet I$ represents the adjustment costs (in terms of lost output). Adjustment costs are further described later.

The model also includes an exogenously specified rate of technological change. We represent such change as Harrod-neutral (labor-embodied). For any given mix of productive inputs (labor, capital, energy, and materials), the productivity of the labor input increases as a function of time. Thus, effective hours worked are actual hours adjusted for the rate of technological change. This rate of change determines the growth of all variables on the steady-state balanced growth path.

Oil Production

In all industries other than the crude oil extraction industry, production exhibits constant returns to scale. In the crude oil industry, in contrast, production involves decreasing returns to scale because of stock effects: as reserves are depleted, producers need to introduce more costly extraction methods in existing fields (e.g., introduce secondary recovery techniques to replace primary recovery extraction methods) or move to new fields that are less fecund.[2] In either case, the depletion of reserves is accompanied by an increase in unit costs. In making current decisions about rates of extraction, forward-looking producers in this industry take into account the impact of their actions on future costs.

In industries other than crude oil extraction, the value-added productivity parameter γ_f is fixed over time. In the crude oil extraction industry, this parameter is a decreasing function of cumulative extraction.[3] Productivity in any given period t is given by

$$\gamma_{f,\text{oil},t} = \gamma_{f,\text{oil},0}\left[1-(Q_t/\bar{Q})^\varepsilon\right] \qquad (\text{B.10})$$

where $\gamma_{f,\text{oil},0}$ represents productivity for crude oil extraction in the benchmark year, Q_t represents cumulative extraction at the beginning of year t, $\bar{Q}$ represents the total stock of recoverable reserves in the benchmark year, and ε is a curvature parameter that captures the rate of decline in productivity. Cumulative extraction follows a basic equation of motion $Q_{t+1} = Q_t + Y_t$. As shown later, crude oil producers take into account the impact of current extraction on future productivity when making production decisions.

Because marginal extraction costs are rising over time, at some point further extraction of the resource is no longer economically profitable and crude oil producers will choose not to extract the total stock of recoverable reserves; that is, $\lim_{t\to\infty} Q_t < \bar{Q}$.

As production of crude oil diminishes over time, consumption of crude oil is replaced by consumption of a "backstop" supply of oil. This backstop is not economically competitive with the traditional source of oil initially, $\gamma_{f,bs}^{1/\rho_{f,\text{oil}}} < \gamma_{f,\text{oil},0}^{1/\rho_{f,\text{oil}}}$, but over time becomes competitive as the costs of supplying the fossil fuel increase. We assume that the backstop industry is otherwise similar to the crude oil extraction industry (involving, in particular, similar inputs to production). The initial year of backstop production and the productivity of the backstop, $\gamma_{f,bs}$, are reference case assumptions (described in chapter 4).

Investment

In each period, managers invest in structures and equipment, taking into account the adjustment costs associated with the installation or removal of physical capital.

Adjustment costs imply that capital is imperfectly mobile across sectors. This is important because it allows the model to capture how policies differentially affect capital productivity and profits in various industries. Perfect mobility would imply that in response to a new policy, capital instantly is reallocated and marginal products of capital immediately are brought to equality across all industries. This would imply similar profit impacts throughout the economy. The adjustment costs in equation (B.9) earlier are given by the product $\phi(I/K) \bullet I$, where ϕ is a convex function of the rate of investment

$$\phi(I^m/K^m) = \frac{(\xi^m/2)(I^m/K^m - \delta^m)^2}{I^m/K^m} \qquad m \in (s,e) \qquad \text{(B.11)}$$

and where δ^m is the rate of economic depreciation for capital stock m and ξ^m is the marginal cost of adjustment. The law of motion for each capital stock m is given by

$$K_{t+1}^m = I_t^m + (1 - \delta^m) K_t^m \qquad \text{(B.12)}$$

In what follows we let the subscript i range over all industries—both energy and nonenergy industries. We will order industries according to $(E_1, \ldots, E_{N_e}, M_1, \ldots, M_{N_m})$, where N_e and N_m indicate the number of energy and materials industries, respectively. For each industry i, let p_i^d and p_i^f denote the producer prices for domestic and foreign industry goods. The exchange rate e converts the foreign currency to the domestic currency so that the price of foreign good i to domestic agents is p_i^f/e.

New capital goods I_t^s and I_t^e are produced by combining domestic and foreign producer goods, using a fixed Leontief aggregation function, that is, based on fixed coefficients. Let $\text{PFI}_{i,t}^d$ and $\text{PFI}_{i,t}^f$ denote, respectively, total domestic and foreign private fixed investment spending on producer good i, and let $\overline{I}_t^m = \sum_{i=1}^N I_{i,t}^m$ denote the total amount of private investment of capital stock m. The total amount of investment and the total amount of domestic and foreign private fixed investment spending are linked through a Leontief aggregation function:

$$\text{PFI}_{i,t}^d = \alpha_i^{I,s} \overline{I}_t^s + \alpha_i^{I,e} \overline{I}_t^e$$
$$\text{PFI}_{i,t}^f = \alpha_i^{If,s} \overline{I}_t^s + \alpha_i^{If,e} \overline{I}_t^e \qquad \text{(B.13)}$$

where $\sum_{i=1}^{N}[\alpha_i^{I,m}+\alpha_i^{If,m}]=1$ for $m \in \{s,e\}$. The weights are fixed over time. Given the aggregation functions, the unit price of a new unit of capital is a weighted average of domestic and foreign producer prices,

$$p_t^{K^m} = \sum_{i=1}^{N}\left[\alpha_i^{I,m}p_{i,t}^d + \alpha_i^{If,m}(p_{i,t}^f/e)\right] \tag{B.14}$$

Profits and the Behavior of Firms

As indicated earlier, the value of the firm is the present value of after-tax profits net of new share issues. Capital income (before corporate taxes), denoted by π^b, is equal to the value of net output less payments to labor, energy, and materials:

$$\pi^b = p^n Y - (1+\tau^P)wL - p^E E - p^M M \tag{B.15}$$

In equation (B.15), $p^n = p^d(1-\tau^x)$ and it denotes the per-unit price of output net of the output tax τ^x; payments to labor and the energy and material input composites are $(1+\tau^P)wL$, $p^E E$ and $p^M M$, respectively.

After-tax profits are equal to after-tax capital income less interest payments on outstanding debt and property tax payments, plus tax deductions:

$$\pi = (1-\tau^a)\pi^b - (1-\tau^h)(r \cdot \text{DEBT}+\text{TPROP}) + \tau^a \delta_D^s K_D^s + \tau^a \delta_D^e K_D^e + LS \tag{B.16}$$

Capital income is taxed at the corporate rate τ^a. Interest payments $r \cdot \text{DEBT}$ are equal to the gross-of-tax interest rate r times the value of the firm's current debt DEBT. Debt in period t is assumed to be a constant fraction b of the current value of capital

$$\text{DEBT}_t = b[p_{t-1}^{K^s}K_t^s + p_{t-1}^{K^e}K_t^e] \tag{B.17}$$

Because capital decisions for period t are the investment decisions of period $t-1$, capital in period t is valued at the purchase price of new capital in period $t-1$. TPROP represents property taxes paid on the current value of the capital stock

$$\text{TPROP}_t = \tau^{pr}[\rho_{t-1}^{K^s}K_t^s + \rho_{t-1}^{K^e}K_t^e] \tag{B.18}$$

where τ^{pr} is the uniform property tax rate on structures and equipment. The value $\tau^h (r \cdot \mathrm{DEBT} + \mathrm{TPROP})$ represents the value of deductions attributed to interest and property tax payments. The deductions are valued at rate $\tau^{h}.$[4] The third and fourth terms on the right-hand side of (B.16) represent current depreciation allowances, where K_D is the current depreciable capital stock basis and δ_D is the tax depreciation rate on capital. The current depreciable capital stock basis for capital stock $m \in (s,e)$ evolves over time according to

$$K^m_{D,t+1} = p_t^{K^m} I_t^m + (1 - \delta_D^m) K^m_{D,t} \tag{B.19}$$

The final term in (B.16) represents lump-sum transfers to firms, a feature of some of the climate policies considered in this text. Under a carbon tax, for example, firms may receive lump-sum payments to offset their carbon tax obligations. Similarly, under a cap-and-trade program, firms may receive some free allowances; firms getting free allowances are in effect receiving a lump-sum payment equal to the value of these allowances.

A cash-flow identity links the sources and uses of revenues by firms:

$$\pi_t + (\mathrm{DEBT}_{t+1} - \mathrm{DEBT}_t) = \mathrm{DIV}_t + \mathrm{IEXP}_t + \mathrm{SR}_t \tag{B.20}$$

The firm receives cash from profits π_t and new debt issue $\mathrm{DEBT}_{t+1} - \mathrm{DEBT}_t$. The firm uses its cash to finance dividend payments DIV_t, investment expenditures IEXP_t, and share repurchases SR_t. Dividends are assumed to be a constant fraction a of after-tax profits plus appreciation in the value of the capital stock minus economic depreciation,

$$\mathrm{DIV}_t = a[\pi_t + (p_t^{K^s} - p_{t-1}^{K^s})K_t^s + (p_t^{K^e} - p_{t-1}^{K^e})K_t^e - \delta^s p_t^{K^s} K_t^s - \delta^e p_t^{K^e} K_t^e] \tag{B.21}$$

Investment expenditures are equal to the total value of current period investment expenditures

$$\mathrm{IEXP}_t = p_t^{K^s} I_t^s + p_t^{K^e} I_t^e \tag{B.22}$$

In equation (B.20), share repurchases SR_t are the residual, making up the difference between sources and uses of revenue. Negative share repurchases are equivalent to share issues and represent a source rather than a use of cash.

Households (stockholders) require firms to offer them a rate of return comparable to the rate of return on owning private or public debt.[5] The after-tax return to stockholders is equal to the after-tax value of dividends plus capital gains (net of share repurchases). This must be equal to the after-tax return of an investment of the same value that earns the safe rate of return (the return on bonds)

$$(1-\tau^e)\text{DIV}_t+(1-\tau^v)(V_{t+1}-V_t+\text{SR}_t)=(1-\tau^b)rV_t \tag{B.23}$$

where τ^e, τ^v, and τ^b represent personal tax rates on dividend income, capital gains, and interest income, respectively. From equation (B.23) we can derive the following expression for the value of the firm:

$$V_t=\sum_{s=t}^{\infty}\left[\frac{1-\tau^e}{1-\tau^v}\text{DIV}_s+\text{SR}_s\right]d_t(s) \tag{B.24}$$

This equation expresses the value of the firm as the discounted sum of after-tax dividends and share repurchases. The discount factor is

$$d_t(s)=\prod_{u=t+1}^{s}\left[1+\frac{(1-\tau^b)r_u}{1-\tau^v}\right]^{-1}.$$

Optimal Input Intensities, Production Levels, and Investment

As mentioned earlier, firm managers choose variable inputs—domestic and foreign energy, domestic and foreign materials, and labor—and the level of investment to minimize costs and maximize the value of the firm. Because the production function is linearly homogeneous in variable inputs, the optimal intensities of these inputs can be solved for independently of the level of investment or output. As a result, each producer's decision problem can be split into two distinct problems: an "inner nest" problem to decide optimal input intensities and an "outer nest" problem for total production and investment.

For any CES function of the form

$$X=\gamma^{\frac{1}{\rho}}\left[\sum_{i=1}^{n}\alpha_i x_i^\rho\right]^{\frac{1}{\rho}} \tag{B.25}$$

the Lagrangian expression for obtaining the composite X at minimum cost is

$$L = \sum_{i=1}^{n} p_i x_i + \lambda \left\{ \gamma^{\frac{1}{p}} \left[\sum_{i=1}^{n} \alpha_i x_i^p \right]^{\frac{1}{p}} - X \right\} \qquad (B.26)$$

given unit prices p_i for input i. The first-order conditions for each good i and j can be combined and summarized as

$$\frac{x_i}{x_j} = \left[\frac{\alpha_j}{\alpha_i} \frac{p_i}{p_j} \right]^{\frac{1}{p-1}} \qquad (B.27)$$

From this expression we can derive the optimal input intensity for input i,

$$\frac{x_i}{X} = [\alpha_i \gamma]^\sigma \left[\frac{p_i}{p} \right]^{-\sigma} \qquad (B.28)$$

and the unit cost function P

$$P = \gamma^{\frac{\sigma}{1-\sigma}} \left[\sum_{i=1}^{n} \alpha_i^\sigma p_i^{1-\sigma} \right]^{\frac{1}{1-\sigma}} \qquad (B.29)$$

Let τ_{ij}^I denote preexisting intermediate input taxes. The price for domestic good i for industry j, denoted by p_{ij}^d, and the price for foreign good i for industry j, denoted by p_{ij}^f, are

$$\begin{aligned} p_{ij}^d &= p_i^d (1 + \tau_{ij}^I) \\ p_{ij}^f &= (p_i^f / e)(1 + \tau_{ij}^I). \end{aligned} \qquad (B.30)$$

The unit cost of labor is expressed by $p_j^l = w(1 + \tau_j^p)$, where w represents the nominal wage and τ_j^p denotes the employer share of the payroll tax. Based on equation (B.28), the prices p_{ij}^d and p_{ij}^f are sufficient to calculate the optimal shares of domestic and foreign inputs to energy input E_i or materials input M_i in each industry j. Equation (B.29) then indicates the unit price for each intermediate input i for industry j. Given the unit price of each intermediate input, the same equations can be used to calculate the optimal share of energy inputs E_i, $i = 1, ..., N_e$, and the optimal share of material inputs M_i, $i = N_e + 1, ..., N_m$, in their respective composites E and M; as well as the unit prices of these composites. With these unit prices, equations (B.28) and (B.29) solve for the optimal shares of the energy and materials composites in the total intermediate inputs and the unit price of total intermediate inputs. Finally, equations (B.28) and (B.29) solve for the optimal share of total intermediate inputs and labor to total variable inputs and the unit price of variable inputs.

For simplicity, let p^V denote the unit price of variable inputs, and let $\bar{g}$ denote the quantity of the variable input composite. For all non–crude oil extraction industries, the problem of choosing optimal production and investment in period t can be summarized with the following Lagrangian

$$
\begin{aligned}
L_t = \sum_{s=t}^{\infty} \Big[& \alpha^1 p_s^n f(K_s, \bar{g}_s) - \alpha^1 p_s^V \\
& + \sum_{m \in s,e} \{ (\alpha_m^2 p_s^{K^m} - \alpha^3 p_{s-1}^{K^m}) - (\alpha_i^4 p_s^{K^m} + \alpha^1 p_s^n \phi_s^m) I_s^m \} \Big] d_t s \qquad \text{(B.31)} \\
& + \sum_{m \in s,e} \Big\{ B_t^m + \sum_{s=t}^{\infty} \lambda_s^m [I_s^m + (1-\delta^m) K_s^m - K_{s+1}^m] d_t(s) \Big\}
\end{aligned}
$$

where

$\alpha^1 = (1-v)(1-\tau^a)$

$\alpha_m^2 = (b-v)(1-\delta^m)$

$\alpha^3 = b - v + \alpha^1 (rb + \tau^{pr})$

$\alpha_m^4 = 1 - b - (1-v)DD_s^m$

$v = a[1 - (1-\tau^e)/(1-\tau^v)]$

DD_s^m = present value of depreciation deductions on \$1 of investment on capital stock m at time s

B_t^m = present value of depreciation deductions attributable to previous investments on capital stock m at time t

Given the current stock of capital, the manager chooses production by deciding on the level of variable inputs $\bar{g}_s$ and optimal investment in the two capital stocks I_s^m to maximize this Lagrangian. The first-order conditions for this problem can be summarized as

$$\frac{\partial L}{\partial \overline{g}_s} : p_s^n f_{\overline{g},s} = p_s^V \tag{B.32}$$

$$\frac{\partial L}{\partial I_s^m} : \lambda_s^m = \alpha_m^4 p_s^{K^m} + \alpha^1 p_s^n \left[\phi_s^m + \phi_s^{\prime m} \frac{I_s^m}{K_s^m} \right] \tag{B.33}$$

$$\frac{\partial L}{\partial K_{s+1}^m} : \lambda_s^m \left[1 + \frac{(1-\tau^b)r_{s+1}}{1-\tau^V} \right] = \left[\alpha^1 p_{s+1}^n f_{K^m, s+1} + \alpha_m^2 p_{s+1}^{K^m} - \alpha^3 p_s^{K^m} \right.$$

$$\left. + \alpha^1 p_{s+1}^m \phi_{s+1}^{\prime m} \frac{I_{s+1}^m}{K_{s+1}^m} \right] + \lambda_{s+1}^m (1-\delta^m) \tag{B.34}$$

where $f_{\overline{g},s}$ and $f_{K^m,s}$ denote the derivative with respect to $\overline{g}$ and K^m in period s, respectively, and $\phi_s^{\prime m}$ denotes the derivative of the adjustment cost function with respect to the investment rate in period s.[6]

Equation (B.32) equates the marginal cost of variable inputs to the marginal product of the variable inputs. Equation (B.33) equates the marginal cost of a new unit of capital, inclusive of adjustment costs, with the shadow value of capital λ_s^m. Equation (B.34) is the Euler condition that equates the marginal cost of a new unit of capital with the discounted marginal benefit of a new unit of capital in the following period.

The problem of the crude oil extraction industry must be augmented to include the effect of cumulative extraction on future productivity. Let $f(K, \overline{g}, Q)$ denote the gross production function inclusive of cumulative extraction Q. The Lagrangian for the crude oil extraction industry is

$$L_t = \sum_{s=t}^{\infty} \left[\alpha^1 p_s^n f(K_s, \overline{g}_s, Q_s) - \alpha^1 p_s^V \right.$$

$$+ \sum_{m \in s,e} \{ (\alpha_i^2 p_s^{K^m} - \alpha^3 p_{s-1}^{K^m}) - (\alpha_m^4 p_s^{K^m} + \alpha^1 p_s^n \phi_s^m) I_s^m \} \bigg] d_t(s)$$

$$+ \sum_{m \in s,e} \left[B_t^m + \sum_{s=t}^{\infty} \lambda_s^m [I_s^m + (1-\delta^m)K_s^m - K_{s+1}^m] d_t(s) \right] \tag{B.35}$$

$$+ \sum_{s=t}^{\infty} \mu_s \left[f(K_s, \overline{g}_s, Q_s) - \sum_{m \in s,e} \phi_s^m I_s^m + Q_s - Q_{s+1} \right] d_t(s)$$

The additional summation term corresponds to the stock effect related to the change in productivity as a function of cumulative extraction. The first-order equations are

$$\frac{\partial L}{\partial \overline{g}_s} : f_{\overline{g},s}\left(\alpha^1 p_s^n + \mu_s\right) = \alpha^1 p_s^V \tag{B.36}$$

$$\frac{\partial L}{\partial I_s^m} : \lambda_s^m = \alpha_m^4 p_s^{K^m} + \left(\alpha^1 p_s^n + \mu_s\right)\left[\phi_s^m + \phi_s'^m \frac{I_s^m}{K_s^m}\right] \tag{B.37}$$

$$\frac{\partial L}{\partial K_{s+1}^m} : \lambda_s^m \left[1 + \frac{(1-\tau^b)r_{s+1}}{1-\tau^v}\right] = \left(\alpha^1 p_{s+1}^n + \mu_s\right) f_{K^m,s+1} + \alpha_m^2 p_{s+1}^{K^m} - \alpha^3 p_s^{K^m}$$

$$+ \left(\alpha^1 p_{s+1}^n + \mu_s\right)\phi_{s+1}'^m \frac{I_{s+1}^m}{K_{s+1}^m}\right] + \lambda_{s+1}^m (1-\delta^m) \tag{B.38}$$

$$\frac{\partial L}{\partial Q_{s+1}} : \mu_s\left[1 + \frac{(1-\tau^b)r_{s+1}}{1-\tau^v}\right] = [\alpha^1 p_{s+1}^n f_{Q,s+1} + \mu_{s+1}(1+f_{Q,s+1})] \tag{B.39}$$

As before, equation (B.36) equates the marginal cost of variable inputs to the marginal product of the variable inputs. The marginal product now includes the shadow value of cumulative extraction: a marginal increase in input use causes an incremental increase in output and a corresponding negative contribution to future production through the resource stock effect. Equation (B.37) equates the marginal cost of a new unit of capital, inclusive of adjustment costs, with the shadow value of capital λ_s^m. Equation (B.38) is the intertemporal Euler condition that equates the marginal cost of a new unit of capital with the discounted marginal benefit of a new unit of capital in the following period. Equation (B.39) is unique to the crude oil extraction industry, representing the intertemporal Euler condition for the shadow value of cumulative extraction μ_s. This equation equates the current shadow value of extraction to the discounted shadow value of extraction in the following period, adjusted for the contribution of future output on productivity and reserves.

The optimal path of the capital stock of the crude oil extraction industry differs considerably from the optimal path in industries with zero stock effect. Ultimately, the capital stocks (and output) of the crude oil extraction industry approach zero. Because of rising extraction costs, it is not economical to deplete the entire stock of reserves. Hence, cumulative extraction asymptotically approaches a total less than the original stock of reserves.

CONSUMER BEHAVIOR

Our modeling of consumer behavior captures several key aspects of consumer choice: the choice between work hours and leisure time, the choice between current consumption and saving for future consumption, and the allocation of current consumption expenditure across various consumer goods and services.

The E3 model specifies a single, representative household to capture consumer behavior. We describe here the household's utility function and its associated choices for labor supply, consumption, and saving.

Utility

The household has constant-relative-risk aversion (CRRA) utility over "full consumption" C. Expected lifetime utility in period t is given by

$$U_t = \sum_{s=t}^{\infty} \beta^{s-t} \frac{1}{1-\sigma} C_s^{1-\sigma} \tag{B.40}$$

where β represents the discount factor and σ is the coefficient of relative risk aversion.[7] Full consumption is a composite of the consumption of goods and services $\bar{C}$ and leisure ℓ:

$$C = \left[\bar{C}^{\frac{\eta-1}{\eta}} + \alpha_\ell^{\frac{1}{\eta}} \ell^{\frac{\eta-1}{\eta}} \right]^{\frac{\eta}{\eta-1}} \tag{B.41}$$

where the parameter η represents the elasticity of substitution between consumption and leisure and α_ℓ is a leisure intensity parameter.

The household's budget constraint is

$$W_{t+1} - W_t = \bar{r}_t W_t + (1-\tau^L) w_t l_t + G_t^T + G_t^C - T_t^L - \bar{p}_t \bar{C}_t \tag{B.42}$$

where W represents financial wealth (the value of equity in firms and the value of private and public bonds) and $\bar{r}$ is the after-tax return on financial wealth. The term $\bar{r}_t W_t$ therefore represents the after-tax capital income of the household, equal to the after-tax value of dividend payments, realized and unrealized capital gains, and private and public interest payments. After-tax labor income is given by the term $(1-\tau^L) w_t l_t$, where τ^L is the tax rate on wages including both the employee payroll tax rate and the wage income tax rate, w_t is the nominal pretax wage rate, and l_t represents aggregate labor supply. Households are endowed with time $\bar{l}_t$. This endowment is allocated to either leisure or labor: $\bar{l}_t = l_t + \ell_t$. Households receive transfers from the government with a value of G_t^T; in some policy cases they also receive a lump-sum carbon tax rebate of G_t^C. Finally, the household must pay lump-sum taxes T_t^L each period. The price of consumption is $\bar{p}_t$ and therefore $\bar{p}_t \bar{C}_t$ is the value of total consumption. The change in household financial wealth $W_{t+1} - W_t$ must equal total after-tax income less the value of consumption. The change in household financial wealth is equal to the change in the value of private and public debt and the change in the value firm equity (which is equal to unrealized capital gains).

The household chooses aggregate consumption $\bar{C}_t$, leisure ℓ_t, and next period's level of financial wealth W_{t+1} in each period.[8] The Lagrangian for the household problem is

$$L^{hh} = \sum_{s=t}^{\infty} \beta^{s-t} \frac{1}{1-\sigma} \left[\left[\bar{C}_s^{\frac{\eta-1}{\eta}} + \alpha_\ell^{\frac{1}{\eta}} \ell_s^{\frac{\eta-1}{\eta}} \right]^{\frac{\eta}{\eta-1}} \right]^{1-\sigma}$$

$$+ \sum_{s=t}^{\infty} \beta^{s-t} \lambda_s^{hh} [(1+\bar{r}_s)W_s + (1-\tau^L)w_s(\bar{l}_s - \ell_s) + G_s^T + G_s^C - T_s^L - W_{s+1} - \bar{p}_s \bar{C}_s]$$

(B.43)

The first-order conditions for the household problem can be summarized as

$$\frac{\partial L^{hh}}{\partial \bar{C}_s} : \lambda_s^{hh} \bar{p}_s = \left[\bar{C}_s^{\frac{\eta-1}{\eta}} + \alpha_\ell^{\frac{1}{\eta}} \ell_s^{\frac{\eta-1}{\eta}} \right]^{\frac{1-\sigma\eta}{(\eta-1)}} \bar{C}_s^{\frac{-1}{\eta}}$$

(B.44)

$$\frac{\partial L^{hh}}{\partial \ell_s} : \lambda_s^{hh}(1-\tau^L)w_s = \left[\bar{C}_s^{\frac{\eta-1}{\eta}} + \alpha_\ell^{\frac{1}{\eta}} \ell_s^{\frac{\eta-1}{\eta}} \right]^{\frac{1-\sigma\eta}{(\eta-1)}} \alpha_\ell^{\frac{1}{\eta}} \ell_s^{\frac{-1}{\eta}}$$

(B.45)

$$\frac{\partial L^{hh}}{\partial W_{s+1}} : \lambda_s^{hh} = \beta(1+\bar{r}_{s+1})\lambda_{s+1}^{hh}$$

(B.46)

Equations (B.44) and (B.45) equate the marginal cost of consumption to the marginal value of consumption and the marginal cost of leisure to the marginal value of leisure. Conditional on the shadow value of consumption λ_s^{hh}, these equations combine to determine the optimal level of consumption and leisure. Equation (B.46) is the Euler condition that equates the discounted shadow value of consumption over time.

Using equations (B.44) and (B.45), the uncompensated and compensated elasticities of substitution can be derived from the utility function parameters η and α_ℓ, the relative price of consumption and leisure, and the fraction of time spent working relative to the total labor endowment. By combining equations (B.44) and (B.45), we can denote leisure as a function of total consumption on goods and services

$$\ell_s = \alpha_\ell \left(\frac{(1-\tau^L)w_s}{\bar{p}_s} \right)^{-\eta} \bar{C}_s \tag{B.47}$$

Inserting (B.47) into the budget constraint and rearranging, we can show that Marshallian supply for total hours worked is

$$l_s = \bar{l}_s - \alpha_\ell \left(\frac{(1-\tau^L)w_s}{\bar{p}_s} \right)^{-\eta} \left[\frac{(1-\tau^L)w_s + (1+\bar{r}_s)W_s - W_{s+1} + G_s^T - T_s^L}{\bar{p}_s \bar{u}_s} \right] \tag{B.48}$$

where $\bar{u}_s = 1 + \alpha_\ell \left(\frac{(1-\tau^L)w_s}{\bar{p}_s} \right)^{1-\eta}$. The uncompensated elasticity ε^u represents the change in labor supply given a change in the after-tax wage rate, holding nonlabor income (and savings) constant. This elasticity is given by

$$\varepsilon_s^u = \frac{\partial l_s}{\partial(1-\tau^L)w_s} \frac{(1-\tau^L)w_s}{l_s} = \frac{\eta}{\bar{u}_s} \frac{1-\omega_s}{\omega_s} - \frac{\bar{u}_s - 1}{\bar{u}_s} \tag{B.49}$$

where $\omega_s = l_s / \bar{l}_s$ denotes the fraction of the labor endowment spent working. The compensated elasticity of substitution ε^c, the change in labor supply given a change in the after-tax wage rate, holding utility (and savings) constant, is related to the uncompensated elasticity through the Slutsky equation such that $\varepsilon^u = \varepsilon^c + \varepsilon^y$ where ε^y denotes the elasticity of labor earnings with respect to nonlabor income $y_s = (1+\bar{r}_s)W_s - W_{s+1} + G_s^T - T_s^L$. The expression for the income elasticity of labor earnings is

$$\varepsilon_s^y = (1-\tau^L)w_s \frac{\partial l_s}{\partial y_s} = -\frac{\bar{u}_s - 1}{\bar{u}_s} \tag{B.50}$$

Therefore the compensated elasticity of labor supply is

$$\varepsilon_s^c = \frac{\eta}{\bar{u}_s} \frac{1-\omega_s}{\omega_s} \tag{B.51}$$

Consumption

The aggregate consumption composite $\bar{C}$ is a nested consumption composite of N_c consumer goods, $\tilde{C}_j (j=1,\ldots,N_c)$. In chapter 3, figure 3.3 displays the nested structure. At the lowest level of the nest, households choose whether to purchase goods (or services) from domestic or foreign suppliers. Total consumption on good j, C_j^P, is a constant-elasticity-of-substitution composite of domestic and foreign-produced goods

$$C_j^P = \gamma_{c_j}^{\frac{1}{\rho_{cj}}} \left[\alpha_{c_j} (C_{D,j}^P)^{\rho_{cj}} + (1-\alpha_{c_j})(C_{F,j}^P)^{\rho_{cj}} \right]^{\frac{1}{\rho_{cj}}} \qquad (B.52)$$

Given the domestic and foreign consumption prices $p_{D,j}^P$ and $p_{F,j}^P$, the optimal price index for the price of the consumption good C_j^P is p_j^P and is calculated using the optimal CES price index in equation (B.29). The domestic and foreign shares of consumption good j, $C_{D,j}^P/C_j^P$ and $C_{F,j}^P/C_j^P$ are calculated using equation (B.28). The prices $p_{D,j}^P$ and $p_{F,j}^P$ are weighted averages of producer prices, as described later.

At the second level of the nest, a Leontief aggregation function combines the consumption good C_j^P and transportation and trade margins into the final consumption good $\tilde{C}_j$. Trade and transportation margins are assumed to be fixed at the benchmark year level; α_j^T represents the share of trade and transport margins for each good j. Consumption of transportation and trade services for good j, C_j^T, as a share of total consumption of good j, is $C_j^T = \alpha_j^T \tilde{C}_j$; consumption of the produced good as a share of total consumption is $C_j^P = (1-\alpha_j^T)\tilde{C}_j$. Given the Leontief structure, the unit price, exclusive of taxes, of good j, $\hat{p}_j$, is a weighted average of the producer price p_j^P and the trade and transportation costs p_j^T.

$$\hat{p}_j = (1-\alpha_j^T)p_j^P + \alpha_j^T p_j^T \qquad (B.53)$$

Consumption spending on goods and services is subject to ad valorem and/or fixed excise taxes and may also earn income tax deductions and/or credits.[9] The after-tax/subsidy price $\tilde{p}$ is a function of the ad valorem tax τ^{ca}, the excise tax τ^{ce}, the income tax deductions s^d, the average personal income tax rate that applies to deductions τ^s, and the level of tax credits s^c:[10]

$$\tilde{p}_j = (1-\tau^s s_j^d)((1+\tau_j^{ca})\hat{p}_j + \tau_j^{ce}) - s_j^c \qquad (B.54)$$

Appendix C describes the data sources used to derive the value of the ad valorem and excise tax rates and the value of tax deductions and credits for each consumer good.

At the top level of the nest, a Cobb-Douglas function (which implies unit elasticity of substitution between all goods) aggregates total consumption of each good into the consumption composite

$$\bar{C} = \prod_{j=1}^{N_c} \tilde{C}_j^{\alpha_j^C} \tag{B.55}$$

where α_j^C represents the spending share for each good j. As shown in chapter 3, $\bar{p}$, the after-tax price of the consumption composite, is

$$\bar{p} = \prod_{j=1}^{N_c} \tilde{p}_j^{\alpha_j^C} \tag{B.56}$$

Cobb-Douglas aggregation functions imply that the household will keep the value of consumption on each good constant over time as a fraction of the total value of consumption such that

$$\tilde{p}_j \tilde{C}_j = \alpha_j^C \bar{p} \bar{C} \tag{B.57}$$

The linear homogeneity of the nested utility function implies that aggregate consumption, leisure, and savings can be solved for independently of the optimal consumption composites. Given $\bar{p}$, the household solves for total consumption $\bar{C}$ by using the first-order conditions described earlier. Then the household uses the optimal shares described here to allocate consumption across goods and to allocate consumption within a good to domestic and foreign-produced goods and mandatory consumption of transportation and trade services used to deliver the final product from producers to the household.

The consumer goods are created from producer goods using a Leontief aggregation function. The matrix IC indicates the intensity of domestic producer goods used to create each domestic consumer good; the matrix has dimension $N \times N_c$ and it is not changed across periods. Therefore, the price of each consumer good is a weighted average of the producer goods used to create the consumer good,

$$p_{D,j}^P = \sum_{i=1}^{N} IC_{i,j} p_i^d \tag{B.58}$$

We employ an analogous matrix IC^f to convert foreign producer prices into foreign consumer prices $p_{F,j}^P$ and another analogous matrix IC^T to convert domestic producer prices into domestic transportation and trade costs.

GOVERNMENT

Federal, state, and local governments are modeled as a single agent. The government levies taxes and issues debt (owned by households). Revenues are used to finance government purchases of goods (including capital goods) and services, public expenditures on labor, and transfers to households.

Government Revenues

Government revenues come from two sources: tax receipts and deficit financing. Real government debt $\bar{D}_t^g$ is assumed to grow exogenously at the steady-state growth rate, and nominal government debt is equal to the real level of debt times the price level, $D_t^g = \bar{p}_t \bar{D}_t^g$. The level of deficit financing in period t is the change in the nominal level of debt between periods t and $t+1$, $D_{t+1}^g - D^g$.

Tax revenue comes from household income taxes, ad valorem and per-unit taxes on consumer goods and services, taxes on firms, border taxes, and revenues from a carbon tax or a cap-and-trade system in which some emissions allowances are introduced via an auction. Taxes on household income are given by

$$
\begin{aligned}
T_t^H = \tau^L l_t + \sum_{i=1}^{N} [\tau^e \text{DIV}_{i,t} + \tau^v (V_{i,t+1} - V_i + \text{SR}_{i,t}) \\
+ \tau^b r_t \cdot \text{DEBT}_{i,t}] + \tau^b r_t D_t^g + T_s^L
\end{aligned}
\tag{B.59}
$$

Taxes on consumer goods and services are

$$
T_t^S = \sum_{j=1}^{N_c} \left[\tilde{C}_{j,t} (\tau_j^{ca} \tilde{p}_{j,t} + \tau_j^{ce}) \right]
\tag{B.60}
$$

The following expression indicates taxes paid by firms:

$$T_t^F = \sum_{i=1}^{N} \left[\tau_i^a \pi_{i,t}^b + \tau_i^p L_{i,t} + \text{TPROP}_{i,t} - \tau_i^h (r_{i,t} \cdot \text{DEBT}_{i,t} + \text{TPROP}_{i,t} \right.$$
$$\left. - \tau_i^a \delta_{D,i}^s K_{D,i,t}^s - \tau_i^a \delta_{D,i}^e K_{D,i,t}^e \right]$$

(B.61)

We let T_t^B represent net border taxes, that is, tariff revenues minus the value of export subsidies; and we use $T_t^{CO_2}$ to represent revenues from a carbon tax or (as applicable) from cap and trade. Tax receipts are offset by tax expenditures in the form of subsidies on tax-favored consumption

$$S_t = \sum_{j=1}^{N_c} \left[\tilde{C}_j \left[(\tau^s s_j^d (1 + \tau_j^{ca}) \hat{p}_{j,t} + \tau_j^{ce}) + s_j^c \right] \right]$$

(B.62)

Total tax receipts net of tax expenditures are

$$T_t = T_t^H + T_t^S + T_t^F + T_t^B + T_t^{CO_2} - S_t$$

(B.63)

Government Expenditures

The government employs capital, labor, and intermediate inputs to produce publicly provided goods and services. The total value of publicly provided goods and services is held constant across policies. This facilitates our analysis of the utility-based welfare impacts of various policies, since publicly provided goods do not directly enter the household's utility function. Figure 3.4 displays the production nest for the production of government services. A Leontief aggregation function combines capital and variable inputs—labor and intermediate inputs—to make government services. The level of government services is assumed to grow at the steady-state growth rate; therefore, the level of the capital stock and the total level of variable inputs must also grow at the steady-state rate. Labor and intermediate inputs of producer goods are jointly combined into the government variable input using a Cobb-Douglas aggregation function such that

$$G_t^v = (L_t^g)^{\alpha_L^G} \prod_{i=1}^{N} (\bar{G}_{i,t}^P)^{\alpha_i^G}$$

(B.64)

where L_t^g denotes the government input of labor, $\bar{G}_{i,t}^p$ denotes the government input of producer good i, and $\alpha_L^G + \sum_{i=1}^{N} \alpha_i^G = 1$. Given the exogenous level of total government variable inputs G_t^v, the optimal levels of government purchases of labor and inputs are

$$w_t L_t^g = \alpha_t^G G_t^v \bar{p}_t^g$$
$$p_{i,t}^d \bar{G}_{i,t}^p = \alpha_i^G G_t^v \bar{p}_t^g \tag{B.65}$$

where the $\bar{p}_t^g = (w_t)^{\alpha_L^G} \prod_{i=1}^{N} (p_{i,t}^d)^{\alpha_i^G}$ represents the government variable price index.

The government adopts levels of investment sufficient to cause government capital stocks to increase at the steady-state growth rate. Investment expenditures each period replace the depreciated government-owned capital and increase the stock to the appropriate level. Let K_g^s and K_g^e denote the government stocks of structures and equipment, then $I_{g,t}^s = (\delta_g^s + g_r) K_{g,t}^s$ and $I_{g,t}^e = (\delta_g^e + g_r) K_{g,t}^e$ where (δ_g^s, δ_g^e) represent the rates of depreciation on public capital and g_r represents the steady-state growth rate. The total value of government expenditures is the sum of government investment expenditures, government variable input expenditures, nominal government transfers to households, and the nominal value of any lump-sum rebates given to households through climate policy:

$$G_t = p_t^{K_g^s} I_{g,t}^s + p_t^{K_g^e} I_{g,t}^e + G_t^v \bar{p}_t^g + G_t^T + G_t^C \tag{B.66}$$

Government spending on investment goods is allocated across producer goods as public fixed investment spending, GFI_p, using a Leontief aggregation function,

$$\text{GFI}_{i,t} = \alpha_i^{I_g^s} I_{g,t}^s + \alpha_i^{I_g^e} I_{g,t}^e \tag{B.67}$$

where $\sum_{i=1}^{N} \alpha_i^{I_g^m} = 1$ for $m \in \{s, e\}$. The composition of government-owned capital differs from that of privately owned capital. (The former includes battleships and fighter planes.) Hence the shares defining the composition public and private investment differ: $\alpha_i^{I_g^m} \neq \alpha_i^{I^m}$. The per-unit price of new government capital is

$$p_t^{K^m} = \sum_{i=1}^{N} [\alpha_i^{I_g^m} p_{i,t}^d] \tag{B.68}$$

Government Budget Constraint and Revenue Neutrality

The government faces the following budget constraint in each period:

$$T_t + D_{t+1}^g - D_t^g = G_t + r_t D_t^g \tag{B.69}$$

That is, total government revenues T_t plus the increase in debt $D_{t+1}^g - D_t^g$ must be sufficient to finance current government spending G_t plus interest payments on the current debt $r_t D_t^g$.

For the initial (or benchmark) dataset, the value of lump-sum taxes paid T^L is calibrated so that the government's budget constraint holds. In the reference case and all climate policy cases, lump-sum taxes are adjusted to ensure that the constraint is satisfied each period.

The policies we explore are revenue-neutral over time: for each policy, the present value of the net revenue generated must be returned (recycled) to the private sector. Under all policies, we maintain the same paths of real government spending, real transfers, and real debt as those in the reference case, though nominal levels may differ as a result of policy-induced changes in the price level. In order to achieve revenue neutrality, either the marginal rates of some existing taxes must be adjusted or lump-sum rebates must be introduced to ensure that the present value of the revenue recycled is equal to the present value of net revenue. In performing policy experiments, we iteratively adjust the marginal rates or the lump-sum rebates until the revenue-neutrality requirement is met.

Generally, the policies involve either one-time, permanent adjustments to marginal tax rates or lump-sum rebates that grow at a constant rate. Although these adjustments are scaled to assure revenue neutrality intertemporally, by themselves they do not guarantee that the government's budget constraint will be satisfied period by period. To assure budget balance each period, we make minor adjustments to lump-sum taxes in each period. Note that these lump-sum tax adjustments are different from any lump-sum rebates that are introduced as a method of revenue-recycling.

The requirement of intertemporal revenue neutrality is mathematically equivalent to the requirement that the present value of these period-by-period adjustments to lump-sum taxes be zero or, equivalently, that the present value of the lump-sum taxes in the policy case be the same as the present value of these taxes in the reference case. When we run the model, to assure that the chosen cuts in prior tax rates or the lump-sum rebates are such as to yield intertemporal revenue neutrality, we compare the present value of lump-sum taxes in the policy case with this present value in the reference case. The required match is expressed by the condition

$$\sum_{t=1}^{\infty} \frac{T_t^L(\text{ref})}{\overline{p}_t(\text{ref})/\overline{p}_0} d(t) = \sum_{t=1}^{\infty} \frac{T_t^L(\text{pol})}{\overline{p}(\text{pol})/\overline{p}_0} d(t) \tag{B.70}$$

where $T_t^L(\text{pol})$ and $T_t^L(\text{ref})$ denote the policy and reference case time paths for adjustments to lump-sum taxes, $\overline{p}_t / \overline{p}_0$ represents the price index for the consumption good, and $d(t)$ is the real reference case interest rate, expressed by $d(t) = \prod_{u=1}^{t} \left[\frac{1 + \overline{r}_u(\text{ref})}{\overline{p}_u(\text{ref}) / \overline{p}_{u-1}(\text{ref})} \right]^{-1}$.

THE FOREIGN ECONOMY

In several respects, the modeling of the rest of the world parallels that for the U.S. economy. As with the U.S. economy, the foreign economy has a representative household and government. The structural details of the foreign household and government mirror those for the U.S.; to avoid repetition, several of the details are not presented here.

Foreign Production

A single representative firm produces all foreign intermediate inputs, consumer goods, and capital goods, with the exception of oil. The foreign industry uses capital and labor to produce foreign goods

$$X^f = f(k(K^{s,f}, K^{e,f}), L^f) \tag{B.71}$$

where both f and k are Cobb-Douglas aggregation functions. The foreign sector faces the same per-unit adjustment cost function as the domestic industries, and net output is given by

$$Y^f = X^f - \phi^{s,f}(I^{s,f}/K^{s,f})I^{s,f} - \phi^{e,f}(I^{e,f}/K^{e,f})I^{e,f} \tag{B.72}$$

The foreign industry is required to use a fixed level of intermediate inputs of good i, IN_i per unit of output. Each intermediate input is a CES aggregate of intermediate inputs produced in the domestic economy IN_i^d and inputs produced in the foreign economy IN_i^f. Given the price of U.S. goods in the foreign economy, $p_i^{fd} = p_i^d(1 - \tau_i^{ex})e$ (where τ_i^{ex} represents a potential per-unit export subsidy on producer good i), and the price of the foreign good in the foreign economy, $p_i^{ff} = p_i^f$; the foreign producer uses equations (B.28) and (B.29) to obtain the minimum cost composite of domestically and foreign produced input i with per-unit cost p_i^{IN}. Omitting the superscript to denote the foreign sector, we can write capital income for the foreign sector is

$$\pi^b = p^n Y - w(1 + \tau^P)L - \sum_{i=1}^{N} p_i^{IN} IN_i X \tag{B.73}$$

The rest of the foreign firm problem is identical to that of the domestic firm. The first-order conditions with respect to capital and investment are identical to equations (B.33) and (B.34). The first-order condition with respect to total variable inputs for the domestic firm is replaced by a first-order condition with respect to labor for the foreign firm. Again omitting the foreign superscript, we can express this by

$$\left(p_s^n - \sum_{i=1}^{N} p_{i,s}^{IN} IN_{i,s} \right) f_{L,s} = (1 + \tau^P)w_s \tag{B.74}$$

Because the single representative foreign firm produces all foreign output with the exception of oil, the price of all non-oil foreign goods must be equal across goods; hence $p_i^f = p^f$ and $p_{F,j}^P = p^f$.

Foreign Oil Production

The foreign oil sector produces oil and sells it to the domestic economy at the world oil price p_t^o (in the domestic currency) in exchange for a fixed basket of domestically (U.S.) produced intermediate inputs. Let X_t^o denote foreign imports of oil, and let IN_i^o denote the level of domestic good i required to be exported to the foreign economy per unit of oil import. The value of oil-related exports is $\sum_{i=1}^{N} p_{i,t}^d \text{IN}_i^o X_t^o$. If the value, in the foreign currency, of foreign oil imports exceeds the value of the oil-related exports, then foreign households receive oil rents:

$$\text{OILRENTS}_t = \left(p_t^o e_t - \sum_{i=1}^{N} p_{i,t}^d e_t \text{IN}_i^o \right) X_t^o.$$

World Oil Price and the Domestic Price of Oil

Domestic oil and foreign oil are perfect substitutes, reflecting the assumption of fully integrated world markets. The nominal world price of oil p_t^o is equal to an exogenously specified real price of oil $p_t^{o,\text{real}}$ times a price index reflecting the price of oil-related exports, $\text{PPI}_{\text{OIL},t} = \dfrac{\sum_{i=1}^{N} p_{i,t}^d \text{IN}_i^o}{\sum_{i=1}^{N} p_{i,0}^d \text{IN}_i^o}$. Crude oil and the backstop fuel are also perfect substitutes in demand. Therefore, the domestic price of oil will be equal to either the world price of oil, adjusted for any oil tariffs (τ^o), or the price of the backstop fuel described earlier. Let p_t^{bs} denote the price sufficient for the backstop industry to supply fully all domestic demand for the input "crude oil." The equilibrium domestic price of oil $p_{\text{oil},t}^d$ must satisfy

$$p_{\text{oil},t}^d = \left[\begin{array}{ll} p_t^o(1+\tau^o) & (p_t^o < p_t^{bs}) \\ p_t^{bs} & (p_t^o > p_t^{bs}) \end{array} \right. \tag{B.75}$$

When p_t^{bs} exceeds the world oil price, adjusted for tariffs, the domestic production is not sufficient to satisfy domestic demand. In this case, imports of foreign oil are the marginal source of supply, and the quantities of imports adjust to balance domestic supply and demand for the "crude oil" good. In contrast, when domestic crude oil and backstop production can meet domestic demand, backstop production represents the marginal source of supply. Under these circumstances, the price p_t^{bs} adjusts each period such that total demand equals total domestic supply.

Trade Balance

The model assumes trade balance: in every period, the total value of imports to the U.S. equals the total value of U.S. exports.

The total value of imports is equal to the total value of imports of intermediate energy and material goods plus the total value of consumption goods plus the value of foreign capital goods plus the value of oil imports:

$$\text{IMPORTS}_t = \sum_{i=1}^{N_e}(p_{E_{i,t}}^f/e_t)E_{i,t}^F + \sum_{i=1}^{N_m}(p_{M_{i,t}}^f/e_t)M_{i,t}^F$$

$$+\sum_{j=1}^{N_c}(p_{F,j,t}^P/e_t)C_{F,j,t}^P + \sum_{i=1}^{N}(p_{i,t}^f/e_t)[\text{PFI}_{i,t}^f] + p_t^o X_t^o \tag{B.76}$$

The total value of exports is equal to the total value of exports of consumption goods plus the total value of exports to the foreign producer for intermediate goods plus the total value of oil-related exports

$$\text{EXPORTS}_t = \sum_{j=1}^{N_c}p_{D,j,t}^P C_{FD,j,t}^P + \sum_{i=1}^{N}p_{i,t}^d \text{IN}_i^d X_t^f + \sum_{i=1}^{N}p_{i,t}^d \text{IN}_i^o X_t^o \tag{B.77}$$

where $C_{FD,j}^P$ represents the foreign consumers demand for domestically produced consumer good j.

CLIMATE POLICIES

The model represents the various climate policies as follows.

Emissions Pricing

The carbon tax and cap-and-trade policies impose prices on emissions. Under the carbon tax policy, the prices are the carbon tax rates that are imposed in each period. Under cap and trade, the prices are emissions allowance prices in each period. Allowance prices are endogenous. In each period, the model solves for the price that causes A_t, the demand for emissions allowances, to match Z_t, the specified limit for aggregate emissions in that period.[11]

We consider carbon tax and cap-and-trade policies with *modified upstream* points of regulation. Industrial users pay taxes equal to the carbon content of the goods they purchase (i.e., fossil fuels) regardless of source (domestic or foreign). Let p^{CO_2} represent the price of emissions (the tax or allowance price per ton of CO_2). For fossil fuel inputs $i \in \{$oil, gas, coal$\}$, industry j faces net-of-tax intermediate input prices of

$$p_{ij}^d = p_i(1+\tau_{ij}^I)+d_j c_{ij} p^{CO_2}$$
$$p_{ij}^f = (p_i^f / e)(1+\tau_{ij}^I)+d_j c_{ij} p^{CO_2}$$
(B.78)

where d_j is a dummy variable equal to 1 if the industry's emissions are covered by the policy, and c_{ij} is the carbon coefficient that converts industry j's input of good i into carbon dioxide emissions. Total covered carbon dioxide emissions for industry j are

$$Z_j = \sum_{i=1}^{N} d_j c_{ij}(E_{ij}^d + E_{ij}^f)$$
(B.79)

Carbon tax revenues, as well as auction revenues under 100 percent auctioning, are equal to the total level of covered emissions times the price of carbon

$$T^{CO_2} = p^{CO_2} \sum_{i=1}^{N} Z_i$$
(B.80)

Some of the carbon tax policies include tax exemptions, and several of the cap-and-trade policies involve free allocation of allowances. Accounting for these elements involves similar modeling elements. Under a carbon tax policy that includes exemptions, α_i^{EXEMPT} represents the portion of firm i's tax payments that are exempted. Under these policies, the value LS in the expression for after-tax profits, equation (B.16), is $LS_i = \alpha_i^{\text{EXEMPT}} p^{CO_2} Z_i$. Under a cap-and-trade program, free allowances are allocated as a proportion of total allowances α_i^{free}. Under these policies, lump-sum payments to firms are $LS_i = \alpha_i^{\text{free}} p^{CO_2} A$. In both cases, the total value of carbon revenues received by the government is reduced by the value of exemptions or free allowances, $T^{CO_2} = p^{CO_2} \sum_{i=1}^{N} Z_i - \sum_{i=1}^{N} LS_i$.

Clean Energy Standard

The Clean Energy Standard enters the model as a constrained cost minimization problem for the electricity transmission and distribution industry. The Lagrangian expression for obtaining the composite X at minimum cost subject to the CES constraint is

$$L = \sum_{i=1}^{n} p_i x_i + \lambda \left\{ \gamma^{\frac{1}{\rho}} \left[\sum_{i=1}^{n} \alpha_i x_i^\rho \right]^{\frac{1}{\rho}} - X \right\} + p^{\text{CES}} \left\{ \frac{\sum_{i=1}^{n} a_i m_i x_i}{\sum_{i=1}^{n} m_i x_i} - \bar{M} \right\} \tag{B.81}$$

where p^{CES} is the shadow value on the CES constraint. The first-order condition for each good i is

$$p_i = \lambda \left[\gamma^{\frac{1}{\rho}} \left[\sum_{i=1}^{n} \alpha_i x_i^\rho \right]^{\frac{1}{\rho}-1} \alpha_i x_i^{\rho-1} \right] + \lambda^{\text{CES}} m_i (a_i - \bar{M}) \tag{B.82}$$

Let $\hat{p}_i = p_i + p^{\text{CES}} m_i (\bar{M} - a_i)$. We can then derive the optimal input intensity

$$\frac{x_i}{X} = [\alpha_i \gamma]^\sigma \left[\frac{\hat{p}_i}{P} \right]^{-\sigma} \tag{B.83}$$

and the unit cost function P

$$P = \gamma^{\frac{\sigma}{1-\sigma}} \left[\sum_{i=1}^{n} \alpha_i^\sigma \hat{p}_i^{1-\sigma} \right]^{\frac{1}{1-\sigma}} \tag{B.84}$$

Inserting equation (B.83) into the binding CES constraint (such that the shadow value is positive), we can show that

$$\sum_{i=1}^{n}\frac{a_{i}m_{i}\alpha_{i}^{\sigma}\hat{p}_{i}^{-\sigma}}{m_{i}\alpha_{i}^{\sigma}\hat{p}_{i}^{-\sigma}}=\bar{M} \tag{B.85}$$

Rearranging, gives

$$\sum_{i=1}^{n}m_{i}(\bar{M}-a_{i})\alpha_{i}^{\sigma}\hat{p}_{i}^{-\sigma}=0 \tag{B.86}$$

This equation can be further manipulated to show that the total value of the shadow taxes and subsidies sums to zero,

$$\sum_{i=1}^{n}p^{CES}m_{i}(\bar{M}-a_{i})x_{i}=0 \tag{B.87}$$

Gasoline Tax

The increased federal gasoline tax enters the model through increases in the ad valorem tax on the consumption good "motor vehicle fuels,"[12]

$$\tilde{p}_{gas}=\hat{p}_{gas}+\tau_{gas}^{ce}+\tau^{gas} \tag{B.88}$$

where τ^{gas} represents the increased federal gasoline tax. The total value of increased government revenues from the gas tax will be $\tau^{gas}\tilde{C}_{gas}$.

EQUILIBRIUM

General equilibrium in the E3 model is attained when supplies equal demands in all markets and periods of time. Since the decisions of households and firms in the present are based on their perfect-foresight expectations about the future, the conditions for equilibrium in the present depend on future conditions. Current equilibrium prices and future equilibrium prices are interdependent, and the algorithm for solving for equilibrium prices must take this interdependence into account.

We solve for the equilibrium prices by structuring the equilibrium problem as a complementary slackness problem. For each equilibrium condition, either the equation holds with equality and the complementary price is positive, or the equation does not hold with equality and the price is zero. In each period, the requirements of equilibrium are as follows:

1. Supply is greater than or equal to demand for all producer goods i.

$$
\begin{aligned}
Y_{i,t} \geq &\sum_{j=1}^{N_e} \mathrm{IC}_{i,j} C_{D,j,t}^P + \sum_{j=1}^{N_e} \mathrm{IC}_{i,j}^T \alpha_j^T \tilde{C}_{j,t} + \mathrm{PFI}_{i,t}^d + \bar{G}_{i,t}^P + \mathrm{GFI}_{i,t} \\
&+ \sum_{j=1}^{N_e} \mathrm{IC}_{i,j} C_{FD,j,t}^P + \mathrm{IN}_i^d X_t^F + \mathrm{IN}_i^o X_t^o \\
&+ \sum_{j=1}^{N} \left[E_{i,j}^d (i \leq N_e) + M_{i,j}^d (i > N_e) \right]
\end{aligned}
\tag{B.89}
$$

On the right-hand side of equation (B.89), the first two terms represent domestic final good consumption demand; the third, fourth, and fifth terms represent private fixed investment, government spending on intermediate goods, and public fixed investment, respectively. The second row represents exports of consumption goods, intermediate inputs, and oil-related exports, respectively. The final row represents total domestic intermediate goods demand. The terms $(i \leq N_e)$ and $(i > N_e)$ are conditional operators equal to 1 if the condition holds and equal to 0 otherwise.

The price p_i^d is complementary to this equilibrium condition. In the oil industry, in all periods such that oil imports are positive, the left-hand side of the equation is replaced by $Y_{oil,t} + Y_{bs,t} + X_t^o$ and oil imports X_t^o must be complementary to the equilibrium condition. In all periods when oil imports are zero, the left-hand side of the equation is $Y_{oil,t} + Y_{bs,t}$ and the backstop price p_t^{bs} is complementary to the equilibrium condition. The shadow values for capital, oil extraction, and consumption must also be consistent with the equilibrium prices that define market clearing.

2. Total labor supply is greater than or equal to total labor demand by firms and government.

$$
l_t \geq \sum_{i=1}^{N} L_{i,t} + L_t^g
\tag{B.90}
$$

The nominal pretax wage is complementary to this equilibrium condition. When the backstop industry is producing, labor demand from the backstop industry must be added to the right-hand side of the equation. In practice, the nominal pretax wage w is the numeraire. In keeping with Walras's Law, one of the demand conditions can be dropped: it is satisfied when the other equilibrium conditions are satisfied. In our model, we drop the condition regarding labor supply.[13]

3. Total private savings is greater than or equal to total demand for household savings, new private and public borrowing, plus the aggregate change in the value of firms (unrealized capital gains).

$$W_{t+1} - W \geq (D_{t+1}^g - D_t^g) + \sum_{i=1}^{N} \left[(\text{DEBT}_{i,t+1} - \text{DEBT}_{i,t}) + (V_{i,t+1} - V_i) \right] \qquad (\text{B.91})$$

The nominal pretax return to savings r_{t+1} is complementary to this equilibrium condition.

4. The total value of imports (supply of foreign currency) is greater than or equal to the total value of exports (demand for foreign currency).

$$\text{IMPORTS}_t \geq \text{EXPORTS}_t \qquad (\text{B.92})$$

The nominal exchange rate that converts the domestic currency into the foreign currency is complementary to this equilibrium condition.

5. The total level of covered emissions is less than or equal to the total level of allowances (cap-and-trade program only).

$$Z_t = \sum_{i=1}^{N} Z_{i,t} \leq A_t \qquad (\text{B.93})$$

The endogenous carbon price is complementary to this equilibrium condition.

Conditions 1–3 must also hold for the foreign economy. Finally, the intertemporal conditions given by equations (B.34), (B.38), (B.39), and (B.46) that equate shadow values of capital, oil extraction, and consumption over time must also hold given the equilibrium market-clearing prices to ensure that the perfect-foresight expectations conditions are satisfied.

DATA, PARAMETERS, AND THE REFERENCE CASE PATH

This appendix describes the data collection process and parameter calibration process in further detail. The outline generally follows the outline for chapter 4 while offering further details as necessary.

FROM PRIMARY DATA TO A SOCIAL ACCOUNTING MATRIX

Deriving Input-Output Matrices

An input-output (IO) matrix indicates how much of commodity i is used in the production of commodity j. It indicates, for example, how much crude oil is used in the production of refined products or how much coal is used in the production of electricity. The BEA Use tables indicate the value of each commodity used by each industry. If industries were associated with only one commodity, then the Use matrix and the IO matrix would be identical. However, individual industries often produce multiple commodities. The BEA Make table indicates the value of commodities produced by each industry. To create a commodity-by-commodity IO matrix, we combine the information in the Use and Make tables. First, we transform the Make table into a Percentage Make table (M) that expresses the share that each commodity represents of each industry's output. We then matrix-multiply the Use table (U) and the Percentage Make table to create the IO table (IO = U ∗ M).

The following two-commodity, two-industry example illustrates the approach. Industries X and Y produce commodities A and B. Industry Y uses 100 units of commodity A and 20 units of commodity B as inputs into production. Industry Z uses 50 units of commodity A and 40 units of commodity B. The Use matrix (U) for this economy is

	Y	Z
A	100	50
B	20	40

Both industries produce goods A and B. Industry Y's production is divided equally between the two commodities, and industry Z's output is composed of 20 percent of commodity A and 80 percent of commodity B. The Percentage Make matrix (M) for this economy is

	A	B
Y	0.5	0.5
Z	0.2	0.8

To calculate the amount of commodity A used in the production of commodity A, we multiply the number of units of good A used by each industry by the percentage that good A comprises in each industry's output, $0.5 * 100 + 0.2 * 50$. Matrix multiplication of the two matrices will calculate the amount of each commodity used in the production of each commodity, IO = U * M,

	A	B
A	60	90
B	18	42

Aggregation and Disaggregation of BEA Data

The BEA's Annual Industry Accounts organize commodities and industries into seventy-one industry groups. We refer to this organization as the summary level of aggregation. These accounts are the source of the IO matrix (constructed as described earlier) and vectors of labor demand, government purchases (intermediate inputs and public fixed investment), personal consumption expenditures, and private fixed investment by industry.[1,2] These data are mapped into the thirty-five E3 industries in the four steps indicated later. For the majority of E3 industries, multiple BEA industries are combined into a single E3 industry. However, in some cases, a single BEA industry must be mapped into multiple E3 industries. Table C.1 summarizes the mapping from BEA summary-level industries to E3 industries. The methods and assumptions are described here in detail.

Step 1: Aggregation

Aggregate the summary-level data into 30 "pre-E3" industries. This aggregation procedure is straightforward. Use vectors are summed across BEA industries into the Pre-E3 industries, and the IO matrix is summed first across columns, to create a 71 by 30 matrix, and then across the rows to create a 30 by 30 matrix.

Step 2: Disaggregation

Disaggregate single BEA industries into multiple E3 industries. This step makes use of the BEA 2007 Benchmark Industry Accounts. The benchmark data cover 389 industry groups. These data are first aggregated into 36 distinct "Benchmark-E3" industries. This dataset is then used to disaggregate the 30 Pre-E3 industries into 36 Disaggregated-E3 industries. In particular, the summary-level BEA data contain "Mining, except oil and gas," "Utilities," "Petroleum and Coal Products," "Federal Government Enterprises," and "State and Local Government Enterprises." The detailed level data have information on "Coal Mining," a subindustry of "Mining, except oil and gas"; "Electric Utilities," "Water Utilities," and "Natural Gas Distribution," subindustries of "Utilities"; "Federal Electric Utilities," a subindustry of "Federal Government Enterprises"; and "State and Local Electric Utilities," a subindustry of "State and Local Government Enterprises."

Consider three Pre-E3 industries A, B, and C and four Benchmark-E3 industries A, B1, B2, and C. Industries A and C are identical across the two definitions, and industries B1 and B2 are subindustries to industry B (B = B1 + B2). The Pre-E3 IO matrix (with only relevant fields filled in)

TABLE C.1 Mapping from BEA Summary Industries to E3 Industries

	BEA SUMMARY INDUSTRY (IOCODE)	E3 INDUSTRY	2007 NAICS CODES
1	Farms (111CA)	Farms, forestry and fishing	1111–1123
2	Forestry, fishing, and related activities (113FF)	Farms, forestry and fishing	113–115
3	Oil and gas extraction (211)	Oil extraction	211
		Natural gas extraction	211
4	Mining, except oil and gas (212)	Coal mining	2121
		Other mining	2122–2123
5	Support activities for mining (213)	Mining support activities	2131
6	Utilities (22)	Electric transmission and distribution	2211
		Coal-fired electricity generation	2211
		Other-fossil electricity generation	2211
		Nonfossil electricity generation	2211
		Natural gas distribution	2212
		Water utilities	2213
7	Construction (23)	Construction	23
8	Wood products (321)	Wood products	321
9	Nonmetallic mineral products (327)	Nonmetallic mineral products	327
10	Primary metals (331)	Primary metals	331
11	Fabricated metal products (332)	Fabricated metal products	332
12	Machinery (333)	Machinery and misc. manufacturing	333
13	Computer and electronic products (334)	Machinery and misc. manufacturing	334
14	Electrical equipment, appliances, and components (335)	Machinery and misc. manufacturing	335
15	Motor vehicles, bodies and trailers, and parts (3361MV)	Motor vehicles	3361–3363
16	Other transportation equipment (3364OT)	Machinery and misc. manufacturing	3364–3369
17	Furniture and related products (337)	Machinery and misc. manufacturing	337
18	Miscellaneous manufacturing (339)	Machinery and misc. manufacturing	339
19	Food and beverage and tobacco products (311FT)	Food and beverage	311–312
20	Textile mills and textile product mills (313TT)	Textile, apparel, leather	313–314

21	Apparel and leather and allied products (315AL)	Textile, apparel, leather	315–316
22	Paper products (322)	Paper and printing	322
23	Printing and related support activities (323)	Paper and printing	323
24	Petroleum and coal products (324)	Petroleum refineries	32411
		Chemicals, plastics, and rubber	32412–32419
25	Chemical products (325)	Chemicals, plastics, and rubber	325
26	Plastic and rubber products (326)	Chemicals, plastics, and rubber	326
27	Wholesale trade (42)	Trade	42
28	Motor vehicle and part dealers (441)	Trade	441
29	Food and beverage stores (445)	Trade	445
30	General merchandise stores (452)	Trade	452
31	Other retail (4A0)	Trade	442,446,451,453
32	Air transportation (481)	Air transportation	481
33	Rail transportation (482)	Railroad transportation	482
34	Water transportation (483)	Water transportation	483
35	Truck transportation (484)	Truck transportation	484
36	Transit and ground passenger transportation (485)	Transit and ground passenger transportation	485
37	Pipeline transportation (486)	Pipeline transportation	486
38	Other transportation and support activities (487OS)	Other transportation and warehousing	487–488, 492
39	Warehousing and storage (493)	Other transportation and warehousing	493
40	Publishing industries, except internet (includes software) (511)	Communication and information	511
41	Motion picture and sound recording industries (512)	Communication and information	512
42	Broadcasting and telecommunications (513)	Communication and information	513
43	Data processing, internet publishing, and other information services (514)	Communication and information	514
44	Federal reserve banks, credit intermediation, and related activities (521CI)	Services	521–522
45	Securities, commodity contracts, and investments (523)	Services	523
46	Insurance carriers and related activities (524)	Services	524
47	Funds, trusts, and other financial vehicles (525)	Services	525

(continued)

TABLE C.1 (*Continued*)

	BEA SUMMARY INDUSTRY (IOCODE)	E3 INDUSTRY	2007 NAICS CODES
48	Housing (HS)	Real estate and owner-occupied housing	531
49	Other real estate (ORE)	Real estate and owner-occupied housing	531
50	Rental and leasing services and lessors of intangible assets (532RL)	Services	532–533
51	Legal services (5411)	Services	5411
52	Computer systems design and related services (5415)	Services	5415
53	Miscellaneous professional, scientific, and technical services (5412OP)	Services	5412–5414,5416–5419
54	Management of companies and enterprises (55)	Services	55
55	Administrative and support services (561)	Services	561
56	Waste management and remediation services (562)	Services	562
57	Educational services (61)	Services	61
58	Ambulatory health care services (621)	Services	621
59	Hospitals (622)	Services	622
60	Nursing and residential care facilities (623)	Services	623
61	Social assistance (624)	Services	624
62	Performing arts, spectator sports, museums, and related activities (711AS)	Services	711–712
63	Amusements, gambling, and recreation industries (713)	Services	713
64	Accommodation (721)	Services	721
65	Food services and drinking places (722)	Services	722
66	Other services, except government (81)	Services	81
67	Federal government enterprises (GFE)	Services	n/a
		Federal electric utilities*	n/a
68	State and local government enterprises (GSLE)	Services	n/a
		State and local electric utilities*	n/a

Note: In the disaggregation process, federal and state/local electric utilities are aggregated into "Electric utilities," which is then disaggregated into the four E3 electricity industries.

and the Benchmark-E3 IO matrix are given the following two matrices, respectively:

PRE-E3 IO MATRIX

	A	B	C
A		w	
B	x	z	y
C		u	

BENCHMARK-E3 IO MATRIX

	A	B1	B2	C
A		W1	W2	
B1	X1	Z11	Z12	Y1
B2	X2	Z21	Z22	Y2
C		U1	U2	

Subindustry shares are derived from the Benchmark-E3 IO matrix ($S_x = X1/(X1+X2)$) and applied to the Pre-E3 IO matrix to create the Disaggregated-E3 IO matrix.

DISAGGREGATED-E3 IO MATRIX

	A	B1	B2	C
A		$S_w w$	$(1-S_w)w$	
B1	$S_x x$	$S_{z11}z$	$S_{z12}z$	$S_y y$
B2	$(1-S_x)x$	$S_{z21}z$	$S_{z22}z$	$(1-S_y)y$
C		$S_u u$	$(1-S_u)u$	

In applying this approach to disaggregate the 30×30 Pre-E3 IO matrix into the 36×36 disaggregated-E3 IO matrix, we assume that subindustry input-output shares are constant between the benchmark year (2007) and the model year (2013). Similarly, when we disaggregate the Use vectors (labor demand, government intermediate inputs, personal consumption expenditures, and private and public fixed investment) into 36×1 vectors, we assume that subindustry shares for Use vector variables are constant between the benchmark year (2007) and the model year (2013).

Step 3: Reaggregation

Aggregate the 36-sector disaggregated-E3 data into 31 "Almost-E3" sectors. To reduce the number of sectors, very disaggregated non-energy-related sectors are aggregated with existing non-energy-related sectors. "Federal Government Enterprise, except Federal Electric Utilities" and "State and Local Government Enterprises, except State and Local Electric Utilities" are aggregated into the Services sector. "Nonrefining Petroleum and Coal Products" are aggregated into "Chemicals, Plastics, and Rubber." Federal and State/Local Electric Utilities are also aggregated into "Electric Utilities."

Step 4: Disaggregate the Oil and Gas and Electric Utilities

The BEA datasets do not provide sufficient information to disaggregate the Almost-E3 industry "Oil and Gas Extraction" into the E3 industries "Crude Oil Extraction" and "Natural Gas Extraction," or to disaggregate the Almost-E3 industry "Electric Utilities" into the E3 industries "Electric Transmission and Distribution," "Coal-Fired Electricity Generation," "Other-Fossil Electricity Generation," and "Nonfossil Electricity Generation." We therefore supplement the BEA data with data from other sources.

According to the Energy Information Administration (EIA), crude oil represented about 69.5 percent of the total value of crude oil and natural gas produced by wells in the United States in 2013. Without further data on how inputs to oil or natural gas extraction differ, we utilize this revenue share to disaggregate the intermediate inputs and labor input from "Oil and Gas Extraction" to "Oil Extraction" and "Gas Extraction." Following EIA data, the petroleum refining industry is the only purchaser of crude oil; therefore, we allocate all purchases of "Oil and Gas Extraction" by petroleum refining to purchases of "Oil Extraction" by petroleum refining. We then allocate all remaining purchases of "Oil and Gas Extraction" by other industries to "Gas Extraction."

The electric power sector IO matrix and Use vectors are disaggregated based on the following:

All purchases of electricity, either as intermediate inputs or final good consumption of electricity, are assigned to "Electric Transmission and Distribution." Industries and households are not allowed to buy electricity directly from the wholesale generators. This is in line with our assumption that wholesale electricity from generators is sold to the transmission and distribution industry and this industry sells retail electricity to industries and households.

By definition, only "Coal-Fired Electricity Generation" purchases "Coal Mining," and only "Other-Fossil Generation" purchases "Natural Gas Extraction," "Natural Gas Distribution," and "Petroleum Refining" (i.e., heating oil) as inputs to production.

Information on annual generation by each type of generator and the average annual price of electricity is used to assign values of inputs of "Coal-Fired Electricity Generation," "Other-Fossil Electricity Generation," and "Nonfossil Electricity Generation" into "Electric Transmission and Distribution" production. The labor demand for "Electric Utilities" is disaggregated into the four sectors using payroll shares for each subindustry from the 2007 Economic Census.

Finally, to allocate nonenergy inputs into the four sectors, we allocate 50 percent, 1 percent, 1 percent, and 48 percent of material inputs for "Electric Utilities" to "Electric Transmission and Distribution," "Coal-Fired Electricity Generation," "Other-Fossil Electricity Generation," and "Nonfossil Electricity Generation," respectively.[3]

Capital Stocks and Returns to Capital

Data on the level of capital stocks are derived from the BEA National Income and Product Account (NIPA) tables entitled, "Current-Cost Net Stock of Private Structures, Equipment, and Intellectual Property Products by Industry (Table 3.1)." Capital stock levels for Equipment and Intellectual Property Products are added to create a single Equipment/IPP stock. (This is often referred to simply as equipment in the E3 model.) In the E3 model, the electric power sector includes government utilities, and the services sector includes other government enterprises. BEA data on the current-cost net stock of government fixed assets (Table 7.1) are used to calculate stocks attributable to government utilities and other government enterprises, and these stocks are aggregated into the applicable E3 industries.

The industry detail in the NIPA tables roughly matches the summary-level industry aggregation in the BEA Make and Use tables. Capital stocks are aggregated into PreK-E3 industries, including "Mining, except Oil and Gas," "Utilities," and "Petroleum and Coal Products." To disaggregate these industries into E3 industries, we use revenue shares from the 2007 Economic Census.

We assume that each subindustry's share of capital is equal to the subindustry's share of revenue in 2007. Using this assumption, "Mining, except Oil and Gas" capital is disaggregated into "Coal Mining" and "Other Mining." "Utilities" capital is disaggregated into "Electric Utilities," "Water Utilities," and "Natural Gas Distribution." "Petroleum and Coal Products" capital is disaggregated into "Petroleum Refining" and "Nonrefining Petroleum and Coal Products," the latter of which is aggregated into E3 industry "Chemicals, Plastics, and Rubber."

The capital stocks for "Oil and Gas Mining" are also disaggregated using revenue shares, but here we apply the 2013 revenue shares derived from the EIA that were used to disaggregate noncapital factor inputs.

To disaggregate the "Electric Utilities" capital stocks into the four E3 electricity generators, we first disaggregate transmission and distribution capital from generator capital, using a capital share parameter derived from the input-output table of Sue Wing (2006). The E3 industry "Electric Transmission and Distribution" is assigned 53.34 percent of the total power sector capital stock (including government-owned capital). The remaining generator capital is allocated in proportion to the overnight installation cost of each type of generation (weighted for multiple generator types within one E3 generation industry). These costs imply 17.61, 9.50, and 19.54 percent of the electric power capital stocks are allocated to "Coal-Fired Electricity Generation," "Other-Fossil Electricity Generation," and "Nonfossil Electricity Generation," respectively.

The returns to capital derive from base-case data on marginal tax rates, tax depreciation rates, debt ratios, payout ratios, interest rates, and capital depreciation rates. From appendix B, we use equation (B.33) to show that shadow value of capital grows at the rate g_p, the nominal growth rate of prices in the steady state. We can then insert equation (B.33) into equation (B.34) to derive the benchmark values for the marginal value of capital $f_{K^m,0}$. For each industry, the capital factor income input in the SAM vector $\bar{K}$ is equal to the capital factor returns times the level of capital stocks less adjustment costs, summed over the two types of capital:

$$\bar{K} = \sum_m \left[f_{K^m,0} K_0^m - \phi^m (I_0^m / K_0^m) \bullet I_0^m \right].$$

Producer and Consumer Goods

The BEA provides a bridge matrix as part of the Annual Industry Accounts. This matrix translates spending on 76 Personal Consumption Expenditure (PCE; as published in NIPA Table 2.4.5 Personal Consumption Expenditure by Type of Product) into spending on BEA Summary-Level commodities. For each PCE good, the matrix displays the total level of spending on each commodity by the purchaser as well as the level received by each commodity producer and the transportation and trade costs associated with the purchase of the commodity.

The creation of the E3 model's bridge matrix proceeds in four steps. First, the PCE goods are aggregated into 24 E3 consumer goods, and the commodities are aggregated into the 30 Pre-E3 industries. This produces two 30 × 24 IC matrices. Table C.2 displays the mapping from 76 PCE goods into the 24 E3 consumption goods. The first matrix shows the total producer value of final good consumption of consumption good j received by industry i. The second shows the total value of transportation and trade value of consumption good j received by industry i.[4] Second, as in the disaggregation of the IO matrix, each matrix is disaggregated into 36 × 24 matrices using the 2007 detailed bridge matrix. Third, industries are reaggregated, and the oil and gas and electric power sectors are disaggregated (consumers do not directly consume wholesale generation or crude oil or gas, so this step is straightforward) to create the final 35 × 24 matrices for producer values and transportation values. Finally, the matrices are converted into the intensity matrices used in the model to translate spending on consumer goods into spending on industry goods.

TABLE C.2 **Mapping from PCE Categories to E3 Consumption Goods**

	PCE CATEGORIES (BY TYPE OF PRODUCT)	E3 CONSUMPTION GOOD
1	New motor vehicles	Motor vehicles
2	Net purchases of used motor vehicles	Motor vehicles
3	Motor vehicle parts and accessories	Motor vehicles
4	Furniture and furnishings	Furnishings and household equipment
5	Household appliances	Furnishings and household equipment
6	Glassware, tableware, and household utensils	Furnishings and household equipment
7	Tools and equipment for house and garden	Furnishings and household equipment
8	Video, audio, photographic, and information processing equipment and media	Recreation
9	Sporting equipment, supplies, guns, and ammunition	Recreation
10	Sports and recreational vehicles	Recreation
11	Recreational books	Recreation
12	Musical instruments	Recreation
13	Jewelry and watches	Clothing
14	Therapeutic appliances and equipment	Health care
15	Educational books	Education
16	Luggage and similar personal items	Recreation
17	Telephone and facsimile equipment	Communication
18	Food and nonalcoholic beverages purchased for off-premises consumption	Food
19	Alcoholic beverages purchased for off-premises consumption	Alcohol
20	Food produced and consumed on farms	Food
21	Women's and girls' clothing	Clothing
22	Men's and boys' clothing	Clothing
23	Children's and infants' clothing	Clothing
24	Other clothing materials and footwear	Clothing
25	Motor vehicle fuels, lubricants, and fluids	Motor vehicle fuels (and lubricants and fluids)
26	Fuel oil and other fuels	Fuel oil and other fuels
27	Pharmaceutical and other medical products	Health care
28	Recreational items	Recreation
29	Household supplies	Furnishings and household equipment
30	Personal care products	Personal care
31	Tobacco	Tobacco
32	Magazines, newspapers, and stationery	Recreation
33	Net expenditures abroad by U.S. residents	Net foreign travel
34	Rental of tenant-occupied nonfarm housing	Housing
35	Imputed rental of owner-occupied nonfarm housing	Housing

TABLE C.2 (*Continued*)

	PCE CATEGORIES (BY TYPE OF PRODUCT)	E3 CONSUMPTION GOOD
36	Rental value of farm dwellings	Housing
37	Group housing	Housing
38	Water supply and sanitation	Water and waste
39	Electricity	Electricity
40	Natural gas	Natural gas
41	Physician services	Health care
42	Dental services	Health care
43	Paramedical services	Health care
44	Hospitals	Health care
45	Nursing homes	Health care
46	Motor vehicle maintenance and repair	Motor vehicles
47	Other motor vehicle services	Motor vehicles
48	Ground transportation	Public ground
49	Air transportation	Air transportation
50	Water transportation	Water transportation
51	Membership clubs, sports centers, parks, theaters, and museums	Recreation
52	Audio-video, photographic, and information processing equipment services	Recreation
53	Gambling	Recreation
54	Other recreational services	Recreation
55	Purchased meals and beverages	Food services and accomodations
56	Food furnished to employees (including military)	Food services and accomodations
57	Accommodations	Food services and accomodations
58	Financial services furnished without payment	Financial services and insurance
59	Financial service charges, fees, and commissions	Financial services and insurance
60	Life insurance	Financial services and insurance
61	Net household insurance	Financial services and insurance
62	Net health insurance	Financial services and insurance
63	Net motor vehicle and other transportation insurance	Financial services and insurance
64	Telecommunication services	Communication
65	Postal and delivery services	Communication
66	Internet access	Communication
67	Higher education	Education
68	Nursery, elementary, and secondary schools	Education
69	Commercial and vocational schools	Education
70	Professional and other services	Other services

(*continued*)

TABLE C.2 *(Continued)*

PCE CATEGORIES (BY TYPE OF PRODUCT)	E3 CONSUMPTION GOOD
71 Personal care and clothing services	Personal care
72 Social services and religious activities	Other services
73 Household maintenance	Other services
74 Foreign travel by U.S. residents	Net foreign travel
75 Less: Expenditures in the United States by nonresidents	Net foreign travel
76 Final consumption expenditures of nonprofit institutions serving households (NPISHs)	Other services

CONSISTENCY PROCEDURES

The consistency procedures scale personal consumption expenditures, exports, private fixed investment, lump-sum tax payments, and the domestic input-output matrix to be consistent with the equilibrium conditions of the balanced growth path model.

Step 1: Scale private fixed investment to be consistent with balanced growth investment. On the balanced growth path, quantities, including the capital stock, must grow at the steady-state growth rate. Using the equation of motion for the capital stocks (equation B.12), we can derive base-case investment from the capital stock and the capital depreciation rates for each industry, $I_{i,0}^m = \left(g_r + \delta^m \right) K_{i,0}^m$. Total investment spending must equal total private fixed investment for each type of capital stock. From the BEA Use tables, we have information on total private fixed investment spending received by industry by capital stock, $\bar{I}_{i,\text{BEA}}^m$, and let $\bar{I}_{\text{BEA}}^m$ denote the BEA total level of domestic private fixed investment spending by capital stock. Using the BEA Import Use table, we can further assign imports to private fixed investment by industry by capital stock $\bar{I}_{i,\text{BEA},F}^m$ and total foreign private fixed investment $\bar{I}_{\text{BEA},F}^m$. The private fixed investment scaling ratio is equal to the sum of investment divided by the sum of total private fixed investment spending: $R_m^I = \sum_{i=1}^N I_{i,0}^m / (\bar{I}_{\text{BEA}}^m + \bar{I}_{\text{BEA},F}^m)$. Therefore, total base-case domestic private fixed investment by commodity is $\text{PFI}_{i,0}^d = R_I^s \bar{I}_{i,\text{BEA}}^s + R_I^e \bar{I}_{i,\text{BEA}}^e$ and total base-case foreign private fixed investment by commodity is $\text{PFI}_{i,0}^f = R_I^s \bar{I}_{i,\text{BEA},F}^s + R_I^e \bar{I}_{i,\text{BEA},F}^e$.

Step 2: Scale government fixed investment to be consistent with government capital stocks and balanced growth. Capital stocks for the government are assumed to grow exogenously, and investment for the government in the base case must match this assumption, $I_{g,0}^{m} = (\delta_{g}^{m} + g_{r})K_{g,0}^{m}$. The BEA Use table provides information on government fixed investment spending received by each industry by capital stock, $\text{GFI}_{i,\text{BEA}}^{m}$. Because government investment spending must equal total government investment receipts, we scale government fixed investment $R_{G}^{m} = I_{g,0}^{m} / \sum_{i=1}^{N} \text{GFI}_{i,\text{BEA}}^{m}$ and $\text{GFI}_{i,0} = R_{G}^{s} \text{GFI}_{i,\text{BEA}}^{s} + R_{G}^{e} \text{GFI}_{i,\text{BEA}}^{e}$.

Step 3: Scale consumption to be consistent with the household budget constraint. The BEA Use table provides information on consumption of domestically produced goods, and the BEA Import Use table provides information on imports that can be allocated to the consumption of foreign-produced goods. Let $\overline{C}_{\text{BEA}}$ denote the value of total consumption spending, net of consumption taxes and consumption subsidies, in the data. Data on labor demand by industry and government are summed into total labor supply l_0, and firm capital stocks, debt ratios, and dividends rates along with government debt determine the total value of household wealth W_0. Tax rates apply to factor incomes and levels of consumption taxes, and consumption subsidies determine total government revenues from households and firms. Combining the government budget constraint and the household budget constraint, we find the scaling ratio R_C such that

$$R_{C}\overline{C}_{\text{BEA}} = (1 - \tau^{L})w_{0}l_{0} + \overline{r}W_{0} - g_{f}W_{0} + G_{0}^{T} + R_{C}(T_{0}^{S} - S_{0}) - T_{0}^{L} \qquad \text{(C.1)}$$

where $(1 - \tau^{L})w_{0}l_{0}$ represents after-tax labor income, $\overline{r}$ represents the after-tax return on wealth (and therefore $\overline{r}W_{0}$ represents after-tax capital income), $g_{f}W_{0}$ represents the change in household wealth assuming a balanced growth path (g_{f} represents the growth rate inclusive of both real growth and nominal growth), G_{0}^{T} represents benchmark levels of government transfers to households (data derived from Economic Report to the President), T_{0}^{S} and S_{0} represent the levels of tax revenues and subsidies from consumption (which also must be scaled to maintain constant tax rates and subsidies), and T_{0}^{L} represents the level of lump-sum taxes required to balance the government budget constraint conditional on tax revenues on factors and consumption, the size of government debt, and government spending.

Step 4: Scale exports such that the total value of exports equals the total value of imports. Given the benchmark value of imports, we scale non-oil-related exports by the ratio R_T such that trade is in balance

$$\text{EXPORTS}_0 = \sum_{j=1}^{N_c} R_T C_{FD,j0}^P + \sum_{i=1}^{N} R_T \text{IN}_i^d X_0^f + \sum_{i=1}^{N} \text{IN}_i^o X_0^o = \text{IMPORTS}_0 \quad \text{(C.2)}$$

Step 5: Adjust the rows and columns of the domestic input-output table $\overline{X}_D$ so that the zero-profit condition holds for every industry. We use the RAS method for adjusting the domestic input-output table such that the total value of expenditures equals the total value of receipts for each industry. RAS is an iterative method of adjustment of rows and columns of the input-output matrix until balance is achieved. The RAS method applied to our BEA input-output matrix requires 102 iterations.

SECONDARY PARAMETER CALIBRATION

Given primary parameters and benchmark data on quantities, we calibrate secondary parameters so that the benchmark year in the model replicates the benchmark-year data. Benchmark-year data are in the form of values—billions of dollars of inputs and outputs. To convert values into quantities, we normalize all benchmark-year producer prices to 1. Hence a unit in the model is the quantity of output worth $1 billion in the benchmark year.

Production Parameters

The production nest consists of constant elasticity of substitution functions. Given the primary parameter σ, the prices for each good (assumed to be 1), and data on input shares x_i/X, we can use the equation for optimal input shares for constant elasticity of substitution functions (equation (B.28) in appendix B) to derive the production share and scale parameters α_i and γ. Inverting equation (B.28), we have

$$\alpha_i \gamma = \left[\frac{p_i}{P} \right] \left[\frac{x_i}{X} \right]^{\frac{1}{\sigma}} \quad \text{(C.3)}$$

where $P = \sum_{i=1}^{N} p_i \left[\dfrac{x_i}{X} \right]$. Applying the constant elasticity of substitution share parameter condition $\sum_{i=1}^{N} \alpha_i = 1$, we can derive the parameters

$$\gamma = \sum_{i=1}^{N} \left[\frac{p_i}{P} \right] \left[\frac{x_i}{X} \right]^{\frac{1}{\sigma}} \tag{C.4}$$

and

$$\alpha_i = \gamma^{-1} \left[\frac{p_i}{P} \right] \left[\frac{x_i}{X} \right]^{\frac{1}{\sigma}} \tag{C.5}$$

Equations (C.4) and (C.5) can be used to derive the constant elasticity of substitution share and scale parameters for all constant elasticity of substitution functions within the production nest. When labor and capital—the factors of production—are inputs into the nest, the prices for each factor are $p^L = (1 + \tau^p)w$ and $p^K = \dfrac{f(K, \bar{g}) - p^V \bar{g}}{K}$, where (as in appendix B) τ^p represents the employer payroll tax rate, w is the pretax wage (assumed to be 1 in the benchmark period), $f(K, \bar{g})$ represents gross output, p^V represents the unit price of variable inputs, $\bar{g}$ represents the quantity of variable inputs, and K represents the aggregate capital stock.

Cobb-Douglas aggregation functions, used for capital aggregation and foreign production, are calibrated using the first-order conditions for optimal inputs in Cobb-Douglas functions. The Cobb-Douglas equation for capital is $K = \gamma_k \prod_{m \in s, e} (K^m)^{\alpha^m}$. Given the benchmark returns to capital $f_{K^m, 0}$ for each capital stock m, the first-order condition for optimal capital input is

$$f_{K^m, 0} = \alpha_m \frac{K_0}{K_0^m} \tag{C.6}$$

Inverting this equation, we obtain

$$\alpha_m = f_{K^m, 0} \frac{K_0^m}{K_0} \tag{C.7}$$

where $K_0 = \sum_{m \in s, e} K_0^m$. Then, the scalar parameter is simply $\gamma_K = \dfrac{K_0}{\prod_{m \in s, e} (K_0^m)^{\alpha^m}}$.

For the foreign production function, let $X^f = \gamma^f (K^f)^{\alpha^f} (L^f)^{1-\alpha^f}$. We can rewrite the first-order condition with respect to labor, equation (B.74), as

$$f_{L,0} = (1 - \alpha^f) \frac{L_0^f}{X_0^f} = \frac{(1 + \tau^P) w^f}{p^{n,f} - \sum\limits_{i=1}^{N} p_i^{IN} IN_i}. \tag{C.8}$$

Given the value of α^f that satisfies equation (C.8), we can then derive the scale parameter $\gamma^f = \dfrac{K_0^f}{(K_0^f)^{\alpha^f} (L_0^f)^{1-\alpha^f}}$.

For each capital stock m, we derive the Leontief aggregation shares from the BEA Use table data on total private fixed investment spending received by industry by capital stock. The coefficients are $\alpha_i^{I^m} = \bar{I}_{i,\mathrm{BEA}}^m / (\bar{I}_{\mathrm{BEA}}^m + \bar{I}_{\mathrm{BEA},F}^m)$ and $\alpha_i^{I^f} = \bar{I}_{i,\mathrm{BEA},F}^m / (\bar{I}_{\mathrm{BEA}}^m + \bar{I}_{\mathrm{BEA},F}^m)$.

Household Parameters

Given unitary prices, we construct the after-tax unit price of consumption goods using our data on consumption taxes and tax-favored consumption subsidies for each good (described in chapter 4),

$$\tilde{p}_{j,0} = (1 - \tau^s s_j^d)(1 + \tau_j^{ca}) + \tau_j^{ce}) - s_j^c \tag{C.9}$$

Using after-tax prices and the BEA data on consumption spending for each consumer good j, $\tilde{C}_{j,0}$, the Cobb-Douglas aggregation function implies constant expenditure shares

$$\alpha_j^C = \frac{\tilde{p}_{j,0} \tilde{C}_{j,0}}{\sum\limits_{j=1}^{N_c} \tilde{p}_{j,0} \tilde{C}_{j,0}} \tag{C.10}$$

The unit price of consumption is $\bar{p}_0 = \prod\limits_{j=1}^{N_c} \tilde{p}_{j,0}^{\alpha_j^c}$ and the total quantity of consumption expenditures is $\bar{C}_0 = \sum\limits_{j=1}^{N_c} \dfrac{\tilde{p}_{j,0} \tilde{C}_{j,0}}{\bar{p}_0}$. To check for consistency in the calibration, it must be the case that the total value of consumption expenditures equals the scaled value of consumption from the consistency procedure, or $\bar{p}_0 \bar{C}_0 = \bar{R}_C \bar{C}_{\mathrm{BEA}}$.

The household utility parameters η and α_ℓ are calibrated to be consistent with data on expenditures of goods and services and leisure.

To determine the level of leisure, we must first assume the fraction of time spent working ω_0. Given the assumption on fraction of time spent working, benchmark levels of leisure are $\ell_0 = l_0 / \omega_0 - l_0$, where l_0 denotes benchmark data on total labor supply (which equals total labor input by industry and government). By combining the definition of the utility term $\bar{u}_0$ with the optimal ratio for leisure to consumption expenditures (equation B.47), $\bar{u}_0 = 1 + \dfrac{\ell_0}{\bar{C}_0} \dfrac{\bar{P}_0}{(1-\tau^L)w_0}$, then following equation (B.51) for the compensated elasticity of labor supply, the elasticity of substitution between consumption and leisure η is given by $\eta = \varepsilon_0^c \bar{u}_0 \dfrac{1-\omega_0}{\omega_0}$, where ε_0^c is the compensated elasticity of substitution, calibrated to be equal to 0.3 as explained in chapter 4. Using equation (B.47), the preference for leisure parameter α_l is given by $\alpha_l = \dfrac{\ell_0}{\bar{C}_0} \left[\dfrac{\bar{P}_0}{(1-\tau^L)w_0} \right]^{-\eta}$. As described in chapter 4, the fraction of time spent working ω_0 is set such that the income elasticity of labor earnings $\varepsilon_0^y = -\dfrac{\bar{u}_0 - 1}{\bar{u}_0}$ is -0.2.

Government Parameters

The parameters in the government Cobb-Douglas aggregation function are derived to be consistent with data on government (federal, state, and local) purchases of labor and intermediate inputs from the BEA Use tables. With unitary prices in the benchmark period, the total quantity of government expenditures on variable inputs is the sum of the quantity of intermediate inputs plus the quantity of labor inputs, or $G_0^v = \sum_{i=1}^{N} \bar{G}_{i,0}^P + L_0^g$.

By assumption of unitary prices, the price index for government variable prices will be 1 ($\bar{P}_0^g = 1$). From equation (B.65), we can then derive the share parameters

$$\alpha_L^G = L_0^g / G_0^v$$
$$\alpha_L^G = \bar{G}_{i,0}^P / G_0^v \tag{C.11}$$

The government investment Leontief aggregation function shares are derived from the BEA Use tables data on public fixed investment received by industry by capital stock. The coefficients are $\alpha_i^{I^m} = \text{GFI}_{i,\text{BEA}}^m / \sum_{i=1}^{N} \text{GFI}_{i,\text{BEA}}^m$.

THE REFERENCE CASE PATH

The reference case path is calibrated to approximate the EIA Annual Energy Outlook (AEO) 2016 forecasts for energy prices, electricity generation shares, and carbon dioxide emissions. AEO forecasts in 2016 are through the year 2040. All changes to E3 parameters are modeled through 2050, assuming linear changes (constant absolute increments) between 2040 and 2050 that are consistent with AEO 2016 forecasts through 2040.

Fossil Fuel Prices

Fossil fuel prices are calibrated to match roughly the trend of price increases between 2013 and 2040 as predicted by AEO 2016.[5] World oil prices in the E3 model are exogenous, and the growth rate in world oil prices is set to match forecasts for prices in 2040. Natural gas and coal prices are endogenous in the E3 model. To model price increases over time, we exogenously impose a linear decrease in total factor productivity ($\gamma_f^{\overline{pf}}$) so that by 2050 natural gas and coal producers are 24 and 16 percent less productive, respectively, than the benchmark level of productivity.

Electricity Production

To match AEO predictions in generation output per unit of fuel input for coal-fired generators and other-fossil generators (primarily natural gas generators), we impose linear changes in total factor productivity in each industry so that by 2050, coal-fired generators are 11 percent less efficient and other-fossil generators are 31 percent more efficient.

Conditional on changes to fossil fuel prices and fossil generation technology productivities over time, we then exogenously increase the efficiency of nonfossil generators to match the 2040 generation share of nonfossil electricity as predicted by AEO 2016 (40 percent generation from nonfossil). The linear increase in productivity between 2013 and 2050 implies a 26 percent increase in nonfossil electricity generation productivity.

Despite the changes in relative input prices and productivities for coal generation and natural gas generation, the E3 model predicts a much higher share of generation from coal relative to AEO 2016 forecasts. To compensate, we impose linear changes in the share parameters for the inputs of coal-fired electricity generation and other-fossil electricity generation in the constant elasticity of substitution aggregation function for the electric transmission and distribution industry. These changes are set so that the resulting shares roughly match forecasts for the relative shares of electricity generated from coal-fired generators and natural-gas-fired generators in this industry. Specifically, we increase the share parameter for other-fossil generators in constant increments of 0.002 and decrease the coal-fired generator share by the same amount. The increment amounts to 0.7 percent of the benchmark other-fossil generation share parameter.

Consumption of Carbon-Intensive Products

The E3 model yields balanced growth in the long run. In particular, real consumption of goods, including carbon-intensive products, ultimately grows at the steady-state growth rate (1.5 percent per year). AEO 2016 forecasts a growth rate of consumption of carbon-intensive products of much less than 1.5 percent per year. To approximate the trends in AEO 2016 forecasts for consumption of carbon-intensive goods between 2013 and 2040, we reduce the share parameters for private consumption (α_j^C) and public consumption (α_i^G) for energy goods such as electricity, natural gas, and refined products (i.e., home heating oil or motor vehicle fuels) at linear rates of 0.8 to 1.4 percent per year. Further, to match aggregate emissions forecasts and to reduce intermediate inputs of carbon-intensive products in nonenergy industries, we reduce the intermediate input share for energy for material industries at a rate of 0.2 percent per year.

THE DISAGGREGATED HOUSEHOLD MODEL: DATA SOURCES, CALIBRATION, AND WELFARE CALCULATIONS

The disaggregated household (DH) model distinguishes the effects of climate policies across households organized into quintiles according to either income or expenditure in the benchmark year (2013). This appendix offers details on the utility maximization problem for the DH model, the process for linking the benchmark data in the E3 and DH models, and the methods for calculating in the DH model the welfare impacts of policy changes.

THE UTILITY MAXIMIZATION PROBLEM

In each quintile, a representative household makes decisions to maximize intertemporal utility. Each household faces the same intertemporal utility maximization problem as the E3 model's representative household. It allocates time between work and leisure, chooses between current consumption and future consumption, and allocates consumption expenditures across various goods and services. The functional form of the utility maximization problem matches the E3 household problem, but the preferences for leisure and consumption differ across quintiles. Preference parameters that differ are the elasticity of substitution between goods and leisure η^q and the leisure intensity parameter α_ℓ^q. The parameter $\alpha_j^{c,q}$, which determines preferences over consumer goods, also differs.

Each quintile's representative household chooses consumption, leisure, and savings to maximize intertemporal utility subject to a budget constraint. Utility is given by

$$\sum_{s=t}^{\infty} \beta^{s-t} \frac{1}{1-\sigma} \left[\left[(\bar{C}_s^q)^{\frac{\eta^q-1}{\eta^q}} + (\alpha_\ell^q)^{\frac{1}{\eta^q}} (\ell_s^q)^{\frac{\eta^q-1}{\eta^q}} \right]^{\frac{\eta^q}{\eta^q-1}} \right]^{1-\sigma} \tag{D.1}$$

and the budget constraint is

$$\bar{p}_s^q \bar{C}_s^q + \bar{w}_s \ell_s^q = \bar{w}_s \bar{l}_s^q + (1+\bar{r}_s)W_s^q + G_s^q + LS_s^q - T_s^q - W_{s+1}^q \tag{D.2}$$

The household takes the price of the aggregate consumption good $\bar{p}_s^q$, the after-tax wage $\bar{w}_s$, and the after-tax return on capital as given. It also takes as given the level of initial wealth W_0, the annual labor endowment $\bar{l}_s^q$, annual government transfers G_s^q, climate policy lump-sum rebates (if applicable) LS_s^q, and lump-sum taxes T_s^q. The Lagrangian for the household problem is

$$L^q = \sum_{s=t}^{\infty} \beta^{s-t} \frac{1}{1-\sigma} \left[\left[(\bar{C}_s^q)^{\frac{\eta^q-1}{\eta^q}} + (\alpha_\ell^q)^{\frac{1}{\eta^q}} (\ell_s^q)^{\frac{\eta^q-1}{\eta^q}} \right]^{\frac{\eta^q}{\eta^q-1}} \right]^{1-\sigma}$$

$$+ \sum_{s=t}^{\infty} \beta^{s-t} \lambda_s^q \left[(1+\bar{r}_s)W_s^q + \bar{w}_s(\bar{l}_s^q - \ell_s^q) + G_s^q + LS_s^q - T_s^q - W_{s+1}^q - \bar{p}_s^q \bar{C}_s^q \right] \tag{D.3}$$

The first-order conditions for the household are

$$\frac{\partial L^q}{\partial \bar{C}_s^q} : \lambda_s^q \bar{p}_s^q = \left[(\bar{C}_s^q)^{\frac{\eta^q-1}{\eta^q}} + (\alpha_\ell^q)^{\frac{1}{\eta^q}} (\ell_s^q)^{\frac{\eta^q-1}{\eta^q}} \right]^{\frac{1-\sigma\eta^q}{(\eta^q-1)}} (\bar{C}_s^q)^{\frac{-1}{\eta^q}} \tag{D.4}$$

$$\frac{\partial L^q}{\partial \ell_s^q} : \lambda_s^q \bar{w}_s = \left[(\bar{C}_s^q)^{\frac{\eta^q-1}{\eta^q}} + (\alpha_\ell^q)^{\frac{1}{\eta^q}} (\ell_s^q)^{\frac{\eta^q-1}{\eta^q}} \right]^{\frac{1-\sigma\eta^q}{(\eta^q-1)}} (\alpha_\ell^q)^{\frac{1}{\eta^q}} (\ell_s^q)^{\frac{-1}{\eta^q}} \tag{D.5}$$

$$\frac{\partial L^q}{\partial W_{s+1}^q} : \lambda_s^q = \beta(1+\bar{r}_{s+1})\lambda_{s+1}^q \tag{D.6}$$

These first-order conditions are equivalent to those of the (single) representative household in the E3 model.[1] As in E3, households have Cobb-Douglas preferences over consumption goods and services. The share parameters, $\alpha_j^{C,q}$, are constant over time. Therefore, the price of the aggregate consumption good is given by

$$\bar{p}^q = \prod_{j=1}^{N_c} \tilde{p}_j^{\alpha_j^{C,q}} \tag{D.7}$$

where $\tilde{p}_j$ represents the net-of-tax purchase price of consumer good j.

To link the E3 and disaggregated household models, we use prices for consumption goods, the after-tax wage, and the after-tax return to capital from the E3 model's reference case and policy case simulations.

Linking the Benchmark E3 Data with the Disaggregated Household Data

We allocate the E3 model data so that the sum of the benchmark values across quintiles is consistent with the aggregate E3 data. The E3 benchmark values for the labor endowment, benchmark wealth, government transfers, and lump sum taxes are allocated to quintiles based on each quintile's share of aggregate labor, capital, transfer income, and lump-sum taxes paid. In what follows, we will use the superscript q to denote value for a given quintile q.

Let α_K^q denote the share of capital income for quintile q. The level of financial wealth for each household is

$$W_0^q = \alpha_K^q W_0 \tag{D.8}$$

where W_0 denotes the benchmark level of financial wealth in the E3 model (equal to the total value of firms plus the total level of public and private debt held by the household).

Let α_L^q denote the share of labor income for quintile q. The level of labor supply for each household is

$$\bar{l}_0^q - \ell_0^q = \alpha_L^q(\bar{l}_0 - \ell_0) \tag{D.9}$$

where $\bar{l}_0 - \ell_0$ denotes the benchmark level of labor supply in the E3 model. To set the total time allocation $\bar{l}_0^q$ and leisure ℓ_0^q, we assume that each quintile spends the same fraction of time working as opposed to enjoying leisure.

Let α_T^q denote the share of transfer income for quintile q. The level of transfer income for each household is

$$G_0^q = \alpha_T^q G_0^T \tag{D.10}$$

where G_0^T denotes the benchmark level of government transfers in the E3 model.

The household's budget constraint is given by

$$\bar{p}_0^q \bar{C}_0^q + \bar{w}_0 \ell_0^q = \bar{w}_0 \bar{l}_0^q + \bar{r} W_0^q + G_0^q - T_0^q - ((1 + g_f) W_0^q - W_0^q) \tag{D.11}$$

The far-right term in equation (D.11) is the household's saving. We assume that this saving is consistent with the balanced growth path. The level of wealth in the following period must equal current period wealth times the full growth rate.[2]

The steps just described yield all of the elements of the budget constraint except for the quintile's lump-sum taxes T_0^q. We set the level of lump-sum taxes to satisfy the budget constraint in equation (D.11). Given the level of lump-sum taxes for each household and the benchmark level of lump-sum taxes in the E3 model T_0^L, we set the share of lump-sum taxes α_{LS}^q such that

$$\alpha_{LS}^q = T_0^q / T_0^L \tag{D.12}$$

To calibrate the utility function parameters η^q, α_ℓ^q, and $\alpha_j^{C,q}$, we use benchmark levels of prices from the E3 model and the benchmark level of quantities for each household to determine the utility-consistent parameter values. For each household, we have data on the level of consumption expenditure by good $\tilde{p}_j \tilde{C}_j^q$, and given the Cobb-Douglas aggregation function, the consumption good share parameters are $\alpha_j^{C,q} = \dfrac{\tilde{p}_j \tilde{C}_j^q}{\sum\limits_i \tilde{p}_i \tilde{C}_i^q}$.

For parameters η^q and α_ℓ^q, we jointly solve for the parameters that solve the following equations:

$$\ell_0^q = \alpha_\ell^q \left(\frac{\bar{w}_0}{\bar{p}_0^q} \right)^{-\eta^q} \bar{C}_0^q \tag{D.13}$$

$$\varepsilon_0^c = \frac{\eta^q}{\bar{u}^q} \frac{1 - \omega_0}{\omega_0} \tag{D.14}$$

$$\bar{u}^q = 1 + \alpha_\ell^q \left(\frac{\bar{w}_0}{\bar{p}_0^q} \right)^{1-\eta^q} \tag{D.15}$$

where equation (D.13) is derived from combining first-order conditions (D.4) and (D.5) and ε_0^c is the benchmark compensated elasticity of labor supply (fixed to the value in the E3 model).

WELFARE CALCULATIONS

Welfare calculations for the E3 representative household and the disaggregated households are identical because of the common structure of the utility maximization problems. In the following text, quintile superscripts have been removed without sacrifice of clarity.

Calculating the Equivalent Variation

First, we define expenditure functions for consumption and leisure. In each case, expenditure is a function of the consumption price, the after-tax wage, and the shadow value of consumption

$$C_t = c(\bar{p}_t, \bar{w}_t, \lambda_t) \tag{D.16}$$

$$\ell_t = \ell(\bar{p}_t, \bar{w}_t, \lambda_t) \tag{D.17}$$

These functions are based on equations (D.4) and (D.5) (or equations (B.44) and (B.45) for the E3 household). Using equilibrium quantities of consumption leisure demand in the policy case, we define intertemporal policy case utility over the periods 0 to S as

$$U_S(\text{pol}) = \sum_{t=0}^{S} \beta^t \frac{1}{1-\sigma} \left[\left[\bar{C}_t(\text{pol})^{\frac{\eta-1}{\eta}} + \alpha_\ell^{\frac{1}{\eta}} \ell_t(\text{pol})^{\frac{\eta-1}{\eta}} \right]^{\frac{\eta}{\eta-1}} \right]^{1-\sigma} \tag{D.18}$$

where $\bar{C}_t(\text{pol}) = c(\bar{p}_t(\text{pol}), \bar{w}_t(\text{pol}), \lambda_t(\text{pol}))$ and $\ell_t(\text{pol}) = \ell(\bar{p}_t(\text{pol}), \bar{w}_t(\text{pol}), \lambda_t(\text{pol}))$.

The equivalent variation is defined as the change in wealth under reference-case conditions that has the same impact on utility as that of the policy change (Mas-Colell, Whinston, and Green 1995). In the E3 model, the equivalent variation welfare problem is solved by finding the sequence of consumption expenditure that is consistent with the household's first-order conditions (D.4–D.6) evaluated at reference case prices that yields a level of utility equal to the policy case level of utility. This problem can be summarized as follows: find the sequence of shadow values of consumption consistent with the intertemporal Euler condition from the household problem (equation (D.6)), terminating in value λ_S^{TC},

$$\begin{bmatrix} \lambda_t(\text{ev}) = \beta(1+\bar{r}_{t+1}(\text{ref}))\lambda_{t+1}(\text{ev}), t < S - 1 \\ \lambda_{S-1}(\text{ev}) = \beta(1+\bar{r}_S(\text{ref}))\lambda_S^{TC} \end{bmatrix} \tag{D.19}$$

and the corresponding optimal choices of consumption and leisure, $\bar{C}_t(\text{ev})=c(\bar{p}_t(\text{ref}),\bar{w}_t(\text{ref}),\lambda_t(\text{ev}))$ and $\ell_t(\text{ev})=\ell(\bar{p}_t(\text{ref}),\bar{w}_t(\text{ref}),\lambda_t(\text{ev}))$, such that utility equals from this stream of consumption and leisure equals policy case utility,

$$U_S(\text{ev}) = \sum_{t=0}^{S}\beta^t \frac{1}{1-\sigma}\left[\left[\bar{C}_t(\text{ev})^{\frac{\eta-1}{\eta}} + \alpha_\ell^{\frac{1}{\eta}}\ell_t(\text{ev})^{\frac{\eta-1}{\eta}}\right]^{\frac{\eta}{\eta-1}}\right]^{1-\sigma} = U_S(\text{pol}) \tag{D.20}$$

Given reference case prices, interest rates, and equation (D.19), the terminal condition (TC) shadow value λ_S^{TC} is sufficient to define consumption and leisure in each period and total utility; therefore, the welfare problem can be simplified into a problem with one equation and one unknown.

Household wealth is equal to the present value of full expenditures. Reference case expenditures are defined as

$$\text{EX}_{S,\text{ref}} = \sum_{t=0}^{S}(\bar{p}_t(\text{ref})\bar{C}_t(\text{ref})+\bar{w}_t(\text{ref})\ell_t(\text{ref}))d_t \tag{D.21}$$

where d_t represents the reference case nominal discount factor

$$d_t = (1+\bar{r}_0(\text{ref}))\prod_{u=0}^{t}[1+\bar{r}_u(\text{ref})]^{-1} \tag{D.22}$$

The equivalent variation level of wealth, evaluated at reference case prices and interest rates, is

$$EX_{S,ev} = \sum_{t=0}^{S}(\bar{p}_t(\text{ref})\bar{C}_t(\text{ev}) + \bar{w}_t(\text{ref})\ell_t(\text{ev}))d_t \qquad (D.23)$$

and the change in wealth equivalent to the impact of the price change in utility, the equivalent variation EV, is

$$EV_S = EX_{S,ev} - EX_{S,ref} \qquad (D.24)$$

The equivalent variation can be decomposed into source-side and use-side impacts. The source-side impact is the utility impact from the change in the nominal value of sources of income (wages, capital income, and transfers), while the use-side impact is the impact from the change in the prices of goods and services (including leisure) consumed.

$$EV_S = \left[\frac{EX_{S,ev}}{EX_{S,pol}}[EX_{S,pol} - EX_{S,ref}]\right] + \left[\frac{EX_{S,ev}}{EX_{S,pol}}EX_{S,ref} - EX_{S,ref}\right] \qquad (D.25)$$

The first term on the right-hand side is the source-side impact; the second term is the use-side impact. Policy case expenditures are evaluated at policy case prices,[3]

$$EX_{S,pol} = \sum_{t=0}^{S}(\bar{p}_t(\text{pol})\bar{C}_t(\text{pol}) + \bar{w}_t(\text{pol})\ell_t(\text{pol}))d_t \qquad (D.26)$$

In the E3 model, we generally report EV impacts only over the infinite horizon ($S=\infty$). In these cases, we solve for a single steady-state level of first-order consistent expenditures and apply the balance growth path adjusted steady-state value for all $t > T$, where T represents the total number of simulation periods.

In Chapter 7's analysis of distributional impacts across households, we have reported the welfare impacts not only over the infinite horizon but over finite intervals as well, that is, for ($S < T$). For policies that do not start in our benchmark year of 2013, we only consider welfare effects beginning from the year in which the policy is announced, and we therefore adjust the periods such that "period 0" refers to the announcement period.

Measuring Use-Side Impacts

In Chapter 7, we measure the use-side impacts three ways: period-by-period, over a finite time interval, and over an infinite time horizon.

Period-by-period use-side impacts can be measured by calculating the equivalent variation from a static utility maximization problem that holds income constant across the reference case and policy cases. Consider the household problem that chooses consumption and leisure subject to a fixed level of wealth W

$$\max_{\bar{C},\ell} \frac{1}{1-\sigma}\left[\left[\bar{C}^{\frac{\eta-1}{\eta}}+\alpha_\ell^{\frac{1}{\eta}}\ell^{\frac{\eta-1}{\eta}}\right]^{\frac{\eta}{\eta-1}}\right]^{1-\sigma} \tag{D.27}$$

subject to the budget constraint

$$\bar{p}\bar{C}+\bar{w}\ell \leq W. \tag{D.28}$$

Combining the first-order conditions with respect to consumption and leisure with the budget constraint, the Marshallian demands for consumption and leisure are

$$\bar{C}=\bar{p}^{-\eta}P^{\eta-1}W \tag{D.29}$$

$$\ell=\alpha_\ell\bar{w}^{-\eta}P^{\eta-1}W \tag{D.30}$$

where $P=[\bar{p}^{1-\eta}+\alpha_\ell\bar{w}^{1-\eta}]^{\frac{1}{1-\eta}}$ is the unit price of full consumption, $C=\left[\bar{C}^{\frac{\eta-1}{\eta}}+\alpha_\ell^{\frac{1}{\eta}}\ell^{\frac{\eta-1}{\eta}}\right]^{\frac{\eta}{\eta-1}}$, and $PC=\bar{p}\bar{C}+\bar{w}\ell$. Using the definition of P and utility $U=\frac{1}{1-\sigma}C^{1-\sigma}$, we can derive the expenditure function EX that shows the level of expenditures that deliver a given level of utility given prices for consumption and leisure

$$EX(\bar{p},\bar{w},U)=(1-\sigma)U^{\frac{1}{1-\sigma}}[\bar{p}^{1-\eta}+\alpha_\ell\bar{w}^{1-\eta}]^{\frac{1}{1-\eta}} \tag{D.31}$$

By duality, the indirect utility function is

$$V(\bar{p},\bar{w},W)=\frac{1}{1-\sigma}\left[W\left[\bar{p}^{1-\eta}+\alpha_\ell\bar{w}^{1-\eta}\right]^{\frac{-1}{1-\eta}}\right]^{1-\sigma} \tag{D.32}$$

By definition, the single-period equivalent variation of this problem is

$$EX(\bar{p}(ref),\bar{w}(ref),U(pol)) - EX(\bar{p}(ref),\bar{w}(ref),U(ref)) \tag{D.33}$$

Because the level of wealth is fixed in this static household problem, total expenditures are fixed across the reference and policy cases; we can therefore derive the following expression for equivalent variation

$$\frac{P_{\text{ref}}}{P_{\text{pol}}} \text{EX}_{\text{ref}} - \text{EX}_{\text{ref}} \tag{D.34}$$

where $\text{EX}_{\text{ref}} = \text{EX}(\bar{p}(\text{ref}), \bar{w}(\text{ref}), U(\text{ref})) = W$. This equivalent variation is equal to the value of the use-side impacts, US_t, in equation (7.4).

To calculate the use-side impacts over an interval of time, we apply the second expression in equation (D.25),

$$US_S = \frac{\text{EX}_{S,ev}}{\text{EX}_{S,pol}} \text{EX}_{S,\text{ref}} - \text{EX}_{S,\text{ref}} \tag{D.35}$$

Using the expressions for EX_S, the discounted value of use-side impacts over the periods 0 through S is

$$US_S = \left[\sum_{t=0}^{s} \frac{C_{t,ev} P_{t,\text{ref}}}{C_{t,pol} P_{t,pol}} d_t \right] \text{EX}_{S,\text{ref}} - \text{EX}_{S,\text{ref}} \tag{D.36}$$

Therefore, the discounted value of use-side impacts is nearly equivalent to the present value of the series of period-by-period static equivalent variations. Because households are able to shift full consumption over time, the price composite ratio must be weighted by the ratio of consumption under the welfare experiment and the policy experiment. If full consumption (and utility) were equal period by period, then the expression for discounted use-side impacts would exactly equal the present value of period-by-period impacts denoted in equation (D.34).

Measuring Source-Side Impacts

The static model used to generate period-by-period use-side impacts assumed a fixed-level of wealth, and by definition the source-side impacts are zero. To calculate the intertemporal source-side impacts, we start with the first expression in equation (D.25),

$$SS_S = \frac{\text{EX}_{S,ev}}{\text{EX}_{S,pol}} [\text{EX}_{S,pol} - \text{EX}_{S,\text{ref}}] \tag{D.37}$$

Following the household budget constraints, the present value of expenditures equals the present value of the endowment

$$EX_S = \sum_{t=0}^{S} [\bar{w}_t \bar{l}_t + (1+\bar{r}_t)W_t - W_{t+1} + G_t + LS_t - T_t] d_t \qquad (D.38)$$

The fraction $R_S = \dfrac{EX_{S,ev}}{EX_{S,pol}}$ converts the value of endowments into units comparable to the use-side impact, and the source-side impacts can be decomposed into labor, capital, transfer, lump-sum rebate, and lump-sum tax impacts

$$SS_S = SS_S^L + SS_S^K + SS_S^G + SS_S^{LS} + SS_S^T \qquad (D.39)$$

where

$$SS_S^L = R_S \left[\sum_{t=0}^{S} [\bar{w}_t(pol)\bar{l}_t(pol) - \bar{w}_t(ref)\bar{l}_t(ref)] d_t \right] \qquad (D.40)$$

$$SS_S^K = R^S \left[\sum_{t=0}^{S} [(1+\bar{r}_t(pol))W_t(pol) - (1+\bar{r}_t)(ref)W_t(ref) \right.$$
$$\left. - (W_{t+1}(pol) - W_{t+1}(ref))] d_t \right] \qquad (D.41)$$

$$SS_S^G = R_S \left[\sum_{t=0}^{S} [G_t(pol) - G_t(ref)] d_t \right] \qquad (D.42)$$

$$SS_S^L = R_S \left[\sum_{t=0}^{S} [LS_t(pol)] d_t \right] \quad (D.43)$$

$$SS_S^T = -R_S \left[\sum_{t=0}^{S} [T_t(pol) - T_t(ref)] d_t \right] \qquad (D.44)$$

By assumption in the disaggregated household model, the present value of lump-sum taxes is unchanged in each period such that $SS_S^T = 0$ for each S. In the E3 model, our revenue neutrality definition only guarantees that $SS_\infty^T = 0$.

NOTES

1. INTRODUCTION

1. See Intergovernmental Panel on Climate Change (2014a).
2. A recent study published in *Land Use Policy* (Geisler and Currens, 2017) estimates that by 2060, 1.4 billion people could be displaced by sea level rise.
3. The World Bank estimates that by 2030, 3.6 billion people could be at risk for malaria, including 100 million as a result of climate change. See Hallegate et al. (2016).
4. While the signatories to that agreement do not face mandatory targets, the agreement was a landmark in its breadth: nearly every nation pledged to reduce its greenhouse gas emissions. Only Syria and Nicaragua did not sign on.
5. However, greater breadth does not always yield lower costs, as shown in Chapter 5.
6. These attractions notwithstanding, the introduction of a federal policy does not vitiate the case for local efforts. The latter often can serve as a valuable complement to federal-level ones. Regulators at the more local level may have some advantages in addressing particular goals or needs of local residents or businesses.
7. The exceptions stem from petrochemical feedstock and other noncombustion uses of fossil fuels. Noncombustion uses account for less than 6 percent of U.S. fossil fuel use.
8. A renewable portfolio standard (RPS) imposes on utilities the requirement that the share of wholesale electricity deemed "renewable" not fall below a specified level. Thus, it introduces the same type of constraint as the one under the CES except that the category "renewable" for an RPS typically is a bit narrower than the category "clean" that applies to the CES. For example, while some CES policies regard hydroelectric and nuclear power as clean, RPS policies do not apply the category "renewable" to these sources of electricity and typically apply it only to solar, wind, and geothermal power.
9. Leading multiperiod, economywide general equilibrium models that have focused on climate policy include ADAGE, IGEM, USREP, DIEM, and NewERA. Published articles featuring these five models, respectively, are Ross, Fawcett, and Clapp (2009); Jorgenson et al. (2013); Rausch, Metcalf, and Reilly (2011); Ross (2014); and Tuladhar, Mankowski, and Bloomberg (2012).

2. CLIMATE POLICY, FISCAL INTERACTIONS,
AND ECONOMIC OUTCOMES

1. Policy analysts and politicians debate whether climate policies come at a cost. We consider this issue later in this chapter and explore it numerically in later chapters.

2. The same interaction applies to earnings from capital. Higher prices of goods and services lower the real value of interest, dividend, or capital gains income and thereby reduce the real return to owning equity or bonds.

3. In fact, the carbon content per ton of coal can vary somewhat. Anthracite coal has slightly higher carbon content than bituminous coal. We also assume that every ton of coal combusted imposes the same marginal damage.

4. Early articles pointing out this effect include those by Pearce (1991); Oates (1993); and Repetto, Dower, and Geoghegan (1992).

5. Under the CES policy, the revenue-recycling effect can be negative because this policy generates no gross revenue and causes a loss of revenue from existing taxes (a negative tax-base effect). Hence, the net revenue from this policy will be negative. In this case, for the policy to be revenue-neutral after recycling, it must involve an *increase* in the rates of existing taxes.

6. Here we are referring to the "strong" double dividend. The literature also defines a "weak" double dividend. To attain the weak double dividend, the policy's costs need not be zero or negative. What is required is that, in addition to offering an environmental improvement (the first dividend), the policy generate costs that are lower than would have been the case if revenues had been recycled in lump-sum fashion. This often arises when revenues are recycled through cuts in the marginal rates of preexisting taxes. Goulder (1995) offers further analysis of the weak form.

7. The early studies tended to focus on impacts on the labor market, and they employed one-period (static) models. Subsequent studies have considered impacts on capital as well as labor and have addressed dynamic impacts. See, for example, Williams (2002) and Barrage (2015).

8. The difference between primary cost and overall cost has implications for the optimal environmental tax rate. Under conditions of pure competition and in the absence of distortionary taxes, the marginal cost of an environmental tax is the primary cost. Under these conditions, the optimal (net-benefit-maximizing) environmental tax rate is the marginal environmental damage. But in the presence of distortionary taxes, the marginal cost of an environmental tax exceeds primary cost to the extent that the tax-interaction effect dominates the revenue-recycling effect. In this case the marginal cost of an environmental tax exceeds primary cost, and consequently the optimal environmental tax rate is less than the marginal environmental damage. On this issue see, for example, Bovenberg and Goulder (1996); Cremer, Gahvari, and Ladoux (1998); and Williams (2001).

9. Technical analyses are also offered in Parry (1997); Goulder (1998); Parry, Williams, and Goulder (1999); and Parry (2003).

10. In the diagram, the slope of the primary cost line is constant, implying constant marginal abatement costs. This is not critical to the analysis: marginal abatement costs could increase or decrease with abatement.

11. Figure 2.2 offers intuition for this result. Suppose that the (small) increases in prices from the first bit of abatement lower the real wage by a small amount and cause labor supply to decline very slightly from the initial value L_1. Even this small reduction in labor supply creates a first-order (i.e., nonincremental) tax-interaction effect, given the gap between the heights of the S_0 and D curves at L_1.

12. Parry, Williams, and Goulder (1999) estimate a value of about $5.00 per ton of CO_2 for Z, based on data for 1990.

13. See, for example, the general equilibrium simulation analyses in Bovenberg and Goulder (1997) and Jorgenson et al. (2013). The marginal excess burden is a measure of the distortionary cost of a tax. For a given tax, it is the cost to the private sector of a marginal increase in the tax rate, after the revenue is returned lump-sum to the private sector. For example, if a given tax rate is increased just enough to raise one dollar in revenue, and the private sector's cost after returning the revenue lump-sum to the private sector is $1.25, the marginal excess burden is 25 cents. Chapter 4 offers a further discussion.

14. For a detailed analysis of this issue, see Bovenberg and Goulder (1997).

15. Resource rents are revenues to resource owners that exceed the costs of production and a normal profit. Owners of especially accessible or productive resources can earn rents. Owners of oil reserves, for example, will enjoy resource rents when the market price of oil exceeds unit costs of extraction plus a normal per-unit after-tax profit.

16. See Bovenberg and van der Ploeg (1998) and Koskela and Schöb (1999) for analyses of the impacts of revenue-neutral green taxes on employment.

17. See Bento, Jacobsen, and Liu (2015) and Markandya, Gonzalez-Eguino, and Escapa (2013) for analyses of the potential of environmental taxes to generate the double dividend by improving the relative taxation of formal and informal labor.

18. Coulson and Li (2013), for example, estimate that home ownership creates external benefits of approximately $1,300 per household.

19. The nonenvironmental and environmental efficiency of the tax system cannot be cleanly separated in these cases, since there is a direct connection between environmental quality and either the utility from marketed goods or the productivity of labor.

20. See Carbone and Smith (2013) for an analysis of this issue.

21. Heal and Park (2015) find that a warmer climate significantly reduces productivity.

22. See, for example, Hughes, Knittel and Sperling (2008); and Li, Linn, and Muehlegger (2014).

23. See, for example, Boyce and Riddle (2007) and Sedor (2015).

24. See "The Conservative Case for Carbon Dividends" by Baker, Feldstein, Halstead, Mankiw, Paulson, Shultz, Stephenson, and Walton (2017).

25. For example, British Columbia's carbon tax policy involves recycling 25 percent of the revenues via equal lump-sum rebates to each household. Also 23 percent of the revenues is devoted to personal income tax cuts and 50 percent to business tax rate reductions and corporate income tax credits.

3. STRUCTURE OF THE E3 MODEL

1. CGE models have a wide range of applications, including international trade policy, tax reform, economic development, and environmental policy. Their particular features differ, depending on the application in question. From Barro (1993), "The term 'microfoundations' refers to the microeconomic analysis of the behavior of individual agents such as households or firms that underpins a macroeconomic theory."

2. Not all energy inputs are combusted. According to the U.S. Energy Information Administration's *Annual Energy Review 2011*, noncombustion uses represented less than 6 percent of energy consumption on average over the period 1980–2011. Petrochemical feedstocks constituted about one fifth of the noncombustion uses.

3. Our model does not capture certain details related to electricity generation and demand, including regional differences in generator mixes, seasonal and time-of-day demand differences, and electricity transmission constraints. "Bottom-up" models with technological details are an example of models that capture some or all of these details. Examples include Haiku and E4ST.

4. Industry subscripts are omitted for simplicity.

5. Each aggregation function exhibits constant returns to scale. Except in the oil industry, the top-level production function f also exhibits constant returns to scale. As indicated earlier and described further later, production in this industry incorporates stock effects: productivity is a negative function of cumulative extraction and the associated reduction in the remaining resource stock.

6. As discussed in chapter 4, the model is structured so that balanced growth is achieved in the long run, in both the reference (or business-as-usual) case and policy cases. Balanced growth means that relative prices do not change and quantities increase at the same steady-state rate.

7. The extent to which higher fossil fuel costs are passed forward and result in higher input prices depends on the elasticities of output supply. Capital adjustment costs imply that the elasticities are lowest in the short term, when most of the adjustment in capital-output ratios needs to be made. In keeping with the relatively low short-run elasticities of supply, the potential for cost pass-through is more limited in the short run than over the longer term. Over the long term, firms are able to achieve their desired long-run capital intensities. The long-run elasticities of supply are infinite, and in the long run there is 100 percent cost pass-through.

8. See appendix B for details on the expression for after-tax profits, interest payments, levels of debt, property tax payments, and tax deductions.

9. There is no uncertainty or risk in the model. Hence the arbitrage equation abstracts from risk.

10. For a discussion, see, for example, Fagan (1997).

11. In reality, stock effects also apply to natural gas production. We have not modeled stock effects in this industry, however, based on the observation that technological advances in the production of shale gas are expected to offset substantially the negative productivity impacts of stock effects in this industry.

12. See appendix B for a detailed description of first-order conditions.

13. The foreign economy's supply of oil on the world market is assumed to be perfectly elastic at the world oil price.

14. Several studies conclude that the oil market is highly integrated internationally, based on the fact that crude oil of similar grades sells for the same price across regions of the globe. See, for example, Bachmeier and Griffin (2006) and Nordhaus (2009).

15. The price of natural gas, as with the prices of other goods except for crude oil, is endogenous in the model. This is in keeping with the fact that domestic and foreign natural gas markets are not fully integrated, reflecting the inability to cheaply transport large volumes of gas around the globe.

16. In determining the emissions factors to be employed in the model, we account for the fact that feedstock uses of fossil fuels do not yield CO_2 emissions. Chapter 4 offers details.

17. The function implies a unitary income elasticity, which is necessary for balanced growth.

18. Armington trade elasticities are the elasticities of substitution between domestically produced goods and their foreign-produced counterparts.

19. Producer goods are converted into consumer goods using a fixed Leontief matrix. The consumption prices p_j^P and p_i^T are weighted averages of producer prices. Although the weights are fixed over time for each consumption good j, the weights will vary by consumption good given the different producer good mix for each consumption good.

20. The base of an ad valorem tax on a given consumer good is the value of spending on that good, while the base of an excise tax on a good is the quantity consumed of the good. Sales taxes are examples of the former; a gasoline tax is an example of the latter. Tax deductions are valued on the total level of consumption spending, inclusive of taxes, whereas credits are valued on the quantity of goods purchased.

21. Tax deductions reduce adjusted gross income. Hence a dollar spent on a good that is subject to tax deductions lowers taxable income by the rate τ^s, a weighted average of labor income and capital income tax rates. In reality, when consumers are eligible for a tax credit related to the purchase of a given good, the value of the credit does not depend on the price of the good. However, in the model we represent tax credits per dollar of expenditure on a given good. This enables us to take account of the extent to which various tax credits alter relative prices of various consumer goods and thereby affect the allocation of consumer expenditure across commodities.

22. Wealth is equal to the cumulative value of firms, the cumulative value of debt issued by firms, and the value of public debt. The arbitrage condition (equation (3.7)) indicates that the return to these holdings must be equal. Savings is equal to the change in the value of wealth: new debt issue (public and private) plus the change in the value of firms (unrealized capital gains).

23. However, in any particular period, these fixed adjustments usually will not adjust revenue by the exact amount needed to satisfy the government's budget constraint. Supplementary (and minor) adjustments to lump-sum taxes are used to ensure that this constraint is satisfied each period. Revenue neutrality is achieved when the present value of these adjustments is zero. Appendix B offers further details on how the model maintains revenue neutrality under climate policies.

24. GAMS is a software program for mathematical programming. PATH is an algorithm for solving simultaneously a set of nonlinear equations or constraints. For more information on mixed complementarity and the PATH solver, interested readers should refer to the *PATH User Guide* (Ferris and Munson 2000), which features an excellent discussion as well as references to economic applications of mixed complementarity programs.

4. DATA, PARAMETERS, AND THE REFERENCE CASE PATH

1. In the E3 model, households own the firms. Thus $\bar{K}$ represents the capital income, net of taxes, that the household receives from its claims on firms' profits.
2. We use the after-redefinition Make and Use tables that redefine secondary production. The BEA redefines secondary production when the input structure of a secondary product for a particular industry differs significantly from the structure of the primary product.
3. The E3 model has thirty-five industries. For a discussion on how the BEA's seventy-one industries are translated into the model's thirty-five industry categories, please refer to appendix C.
4. See appendix C for details on the transformation of Use and Make tables into an input-output table.
5. Specifically, we use tables on current-cost net stock and current-cost depreciation of private structures, private equipment, and intellectual property by industry.
6. We use the 2013 federal plus state average marginal rates. The tax rate on dividends is a weighted average of ordinary dividends and qualified dividends. We use the long-term marginal rate for capital gains.
7. Source: BEA NIPA Table 3.5: Taxes on Production and Imports, State and Local Property Taxes
8. Source: BEA NIPA Table 3.6: Contributions for Government Social Insurance, Employer Contributions, and Employee Contributions
9. Under GDS, nonresidential real property has a recovery period of 39 years. Recovery periods for equipment are generally 3 10 years; hence we apply a 7-year recovery period as an approximation for all equipment. We bound the tax depreciation rate for each industry, imposing the condition that the tax depreciation rate not fall below the actual depreciation rate.
10. Sector definitions from Damodaran Online do not match BEA industry definitions. A bridge algorithm is used to assign values across BEA industries. Debt and payout ratios are capped at 75 percent.
11. Property taxes are not included in $\bar{T}$ because they are implicitly included in the return to capital.
12. The key requirements for income-expenditure balance are given by equations (B.42) and (B.69) in appendix B.

13. The nominal interest rate is $(1+r)*(1+g_p)-1$, where r is the real interest rate and g_p is the nominal growth rate.

14. We are grateful to Peter J. Wilcoxen for providing both the estimates and the methodology to convert translog elasticities to constant-elasticity-of-substitution elasticities.

15. To ensure model stability, we use adjustment cost values of 10 for the crude oil sector. We use a value of 3.5 for the adjustment cost parameter for the backstop sector.

16. See appendix B for a derivation of the compensated elasticity of labor supply.

17. Kimball and Shapiro (2010) offer an excellent discussion. McClelland and Mok (2012) provide a general review of recent labor supply estimates.

18. Feenstra et al. (2014) distinguish between macro trade elasticities: (1) the elasticity of substitution between domestic and an aggregation of foreign-produced goods, and (2) elasticities of substitution between specific goods produced in different foreign countries. The latter are referred to as macro trade elasticities. The macro trade elasticity is most relevant to our model, since our model represents imports as a composite of all foreign-produced goods.

19. In cases where Feenstra et al. did not provide estimates, we use unit trade elasticities. In most cases, these industries experience very little trade flows (such as services).

20. Foreign demand for U.S. goods can depend on both the macro elasticity and the micro elasticity. If U.S. goods became more expensive, a foreign country could import less from the U.S. and more from other countries (reflecting the micro elasticity), or it could reduce its imports entirely and produce more at home (reflecting the macro elasticity). These considerations imply that foreign trade demand for U.S. goods will be more elastic than U.S. trade demand for foreign goods.

21. See "Estimates of Federal Tax Expenditure for Fiscal Years 2014–2018" from the Joint Committee on Taxation and "IRS, Statistics of Income Division, Publication 1304, July 2015" (Table 1.3) from the Internal Revenue Service.

22. Using information from the EIA on production and current recoverable reserves, the production of oil in 2013 was about 7 percent of recoverable reserves. To account for new discoveries and reserves that become recoverable with increased oil prices, we assume a higher level of total reserves.

23. The scale parameter is γ in the constant-elasticity-of-substitution production function $X = \gamma[\alpha K^\rho + (1-\alpha)g^\rho]^{\frac{1}{\rho}}$ where, as defined in chapter 3, X is output, K is a capital composite (consisting of structures and equipment/intellectual property), and g is a composite of variable inputs.

24. EIA's AEO 2016 provides forecasts only to 2040. For the E3 reference case, we assume trends prior to 2040 continue until 2050.

5. TWO APPROACHES TO CARBON DIOXIDE EMISSIONS PRICING: A CARBON TAX AND A CAP-AND-TRADE SYSTEM

1. As indicated in chapter 3, combustion uses account for over 94 percent of the uses of fossil fuels.

2. According to economic theory, for given levels of emissions pricing, under perfectly competitive conditions the economic outcomes are the same no matter where such pricing is introduced along the supply chain. That is, provided that the tax or cap-and-trade system introduces the same price per unit of carbon content, the equilibrium impacts on output, profits, and emissions will be the same no matter where the regulation is introduced in the supply chain. See, for example, Fischer, Kerr, and Toman (1998). The intuition appeals to the forward- and backward-shifting of the costs of emissions pricing. When emissions pricing is imposed upstream, the additional costs are shifted forward to midstream and downstream users of carbon. When such pricing is imposed downstream, additional costs are shifted backward to initial carbon suppliers. However, this equivalence generally will not hold under imperfect competition. See Ganapati, Shapiro, and Walker (2016).

3. For further discussion of the attractions and limitations of emissions pricing as an instrument for climate policy, see Aldy and Stavins (2012) and Goulder and Parry (2008).

4. See, for example, Hoel and Karp (2002).

5. See, for example, Goulder and Parry (2008).

6. As discussed in chapter 6, under special conditions more conventional regulations such as intensity standards can have a cost advantage relative to emissions pricing.

7. Economic theory claims that if firms are competitive and the emissions prices are at the efficient level, the output prices will be at the efficiency-maximizing level as well.

8. The cost-advantage of emissions pricing depends on how the revenues are recycled. Later in this chapter and in subsequent chapters, we consider the quantitative significance of this issue.

9. Lump-sum revenue recycling can take the form of increases in government transfers or decreases in inframarginal income taxes. In Chapter 7's analysis of distributional impacts across households, the differences in interpretation will be important.

10. The MCPF is the cost to the private sector of raising a unit (e.g., a dollar) of public revenue, before accounting for the gain from any public use of this revenue. Most estimates of the MCPF exceed 1, which means that raising public funds includes an efficiency cost, that is, a cost over and above the revenue raised.

11. According to estimates by the Interagency Working Group on the Social Cost of Greenhouse Gases (2016), the SCC will increase by about 1–2 percent per year. The working group recently adopted the slightly more informative abbreviation SC-CO$_2$. In this book we will use the conventional abbreviation SCC. See Nordhaus (2014) for an analysis of SCC concepts as well as a range of estimates.

    Estimates of the SCC derive from integrated assessment models (IAMs), numerical models that consider how emissions of CO$_2$ lead to changes in CO$_2$ concentrations and changes in climate, as well as the impacts of changes in climate on the economy. These models tend to have less detail on the economy. Leading IAM models include DICE (Nordhaus 2014), FUND (Anthoff and Tol 2013), and PAGE (Hope 2011).

12. The model does not capture the burning of coal by households. Information on such coal use is not available, as the EIA no longer collects information on household use of coal. In 2007, the last year this information was collected, these emissions represented 0.012 percent of domestic emissions. The model also does not include emissions from nonenergy uses of fuels and from production processes such as cement and limestone production.

13. See Metcalf (2017) for a discussion of strengths and weaknesses of different choices for the points of regulation of a carbon tax.

14. In the case of recycling through cuts in individual income tax rates, the tax rates on both individual labor and individual capital income are reduced, with the rates lowered in the same proportion.

15. For example, relative to the reference case, emissions in 2030 are reduced by 27.3 percent under the carbon tax with lump-sum recycling, as compared with 27.1, 27.0, and 26.2 percent under recycling via payroll tax cuts, individual income tax cuts, and corporate tax cuts, respectively. Emissions are slightly higher (and emissions reductions slightly smaller) when revenues are used to finance rate cuts because lower tax rates lead to greater economic activity.

16. Windfall impacts would occur and affect policy-generated revenues at the time of an advance announcement.

17. The equivalent variation is the change in wealth under reference case conditions that has the same impact on utility as that of the policy change. A positive value for the equivalent variation implies that the policy improves welfare. Appendix D offers details.

18. Nominal values were converted to real values using an implicit GDP (or GDP component) deflator.

19. An upstream tax will have virtually identical profit impacts as our central case midstream tax because the midstream tax covers all sources of emissions. If the midstream tax covered only a subset of industries, the profit impacts would be limited to industries that are closely linked to the covered industries.

20. The price impacts are similar in the other recycling cases.

21. As mentioned, in policy cases the time-profile for the real price of oil is the same as in the reference case. However, higher nominal oil prices are needed in policy cases to generate the same real prices as in the reference case. Because the carbon tax leads to price increases, the price index used to convert real prices into nominal prices is higher.

22. As indicated in chapter 3, effective hours are actual hours adjusted for an exogenously specified rate of labor-embodied (Harrod-neutral) technological progress—which implies that, over time, the same actual hours of work imply increasing effective hours. Effective hours worked are not the same as jobs. Thus, table 5.5 does not show the impacts on the number of workers employed. To the extent that there are changes in hours worked per individual worker (an intensive margin impact), changes in jobs in each industry (an extensive margin impact) will differ from the E3 model's changes in labor demand. It should also be kept in mind that E3 does not account for involuntary unemployment, since it is a full-employment model.

23. As indicated earlier, imported oil is represented as the marginal source of supply to U.S. purchasers of crude oil. The reduction in demand for crude oil induced by the carbon tax is accomplished entirely through reduced oil imports. Hence the large decrease in import intensity in this industry.

24. The lower domestic prices of fuels can contribute to lower overall fuel prices on the world market. This can prompt higher quantities demanded of such fuels, offsetting the demand-reducing impact of the U.S. carbon tax. This offset has been termed "fuel price leakage."

25. Economic theory indicates that the distortionary costs of a given tax increase more than in proportion to the tax rate. If the carbon tax were introduced in an economy with no prior taxes, then the gross costs would increase more than in proportion to the carbon tax rate. However, in our simulations, the presence of prior taxes implies that a given percentage increase in the carbon tax rate represents a smaller percentage increase in the overall tax on factors. Hence gross costs rise less than in proportion to the increase in the carbon tax.

26. To evaluate the primary cost for a given carbon tax time profile, we calculate the welfare cost for a carbon tax in a counterfactual economy in which there are no preexisting distortionary taxes.

27. Of course, lowering taxes on capital also raises issues of fairness. We explore issues of fairness in chapter 7, when we examine how the benefits and costs impacts of various climate policies are distributed across various groups.

28. Tax-favored consumption affects the revenue-recycling impact of individual income tax cuts, but not payroll or corporate tax cuts or lump-sum rebates. Therefore we only consider individual income tax cuts in figure 5.8.

29. This analysis shows how reductions in individual income taxes can improve efficiency not only by encouraging factor supply but also by improving the allocation of expenditure across goods and services. Reductions in individual income taxes could work toward efficiency improvements through other channels that we have not considered, such as increases in labor effort. Several studies aim to consider broadly the channels through which reduced income taxes might yield efficiency gains. These studies consider a "taxable income elasticity" parameter that is meant to encompass a broad range of potential impacts. See, for example, Feldstein (1995); Parry (1998); Parry and Bento (2000); and Chetty (2009).

30. "Technical Support Document: Technical Update of the Social Cost of Carbon," Interagency Working Group on Social Cost of Greenhouse Gases, U.S. Government, August 2016.

31. The Interagency Working Group does not report values for the SCC after 2050. We assume that the SCC remains at the 2050 value after that year.

32. PM_{10} refers to particles less than 10 micrometers in diameter; $PM_{2.5}$ refers to particles less than 2.5 micrometers in diameter.

33. We also use Air Quality Trends data from the EPA (https://www.epa.gov/air-trends) to calculate the change in criteria pollutant emissions from the transportation and electricity generation sectors between 2011 and 2013.

34. $PM_{2.5}$ is known to cause irregular heartbeat, asthma attacks, reduced lung function, heart attacks and premature deaths. $PM_{2.5}$ concentrations are caused by direct $PM_{2.5}$ emissions as well as emissions of NO_x and SO_2, which are precursors to ambient levels of $PM_{2.5}$. EPA provided the benefit numbers in "Technical Support Document: Estimating the Benefit per Ton of Reducing $PM_{2.5}$ Precursors from 17 Sectors."

35. There is considerable debate as to the appropriate discount rate to apply to future impacts of climate change. For a discussion, see Nordhaus (2007), Stern (2008), Weitzman (2010), and Goulder and Williams (2012).

36. Here we alter parameter values prior to the model's calibration to yield balanced growth. Changing a given parameter at this stage generally necessitates adjustments in other parameters as part of the calibration process. We have also employed an alternative approach in which parameters are altered at the post-calibration stage. This alternative approach yields a very similar pattern of results.

37. Cumulative emissions are measured as the present value of emissions over the infinite horizon, discounted using the time profile of interest rates in the reference case. These are approximately 3 percent. Discounted cumulative emissions are the denominator in the welfare cost per ton reduced calculations in which period-by-period welfare costs divided by the change in emissions are discounted over time. Different parameter values imply different time paths of emissions. As a result, in this context involving different parameter assumptions, policies with the same cumulative emissions reductions will involve different absolute reductions in emissions across periods.

38. CO_2 currently contributes about 81 percent of the emissions of all greenhouse gases, where emissions of different gases are compared based on their estimated "carbon equivalents," that is, their contributions to global warming. The CO_2 emission reductions required to meet the targets depends on reductions in non-CO_2 greenhouse gases and increases in sinks. Chen and Hafstead (2016) show that, in the absence of new regulations on non-CO_2 greenhouse gases and assuming zero changes in land-use sequestration (sinks), carbon dioxide emissions would need to decrease by 33.2 percent relative to 2005 to meet the 28 percent net greenhouse gas reduction target.

39. We use the expressions "given out free" and "freely allocated" rather than "grandfathered" to refer to any case involving the free provision of allowances. Grandfathering is a special case of free allocation, wherein firms are offered allowances matching historical emissions at some prior time. Thus, firms are "grandfathered" the right to continue to emit at historical levels without charge.

40 Our analysis does not take into account potential transaction costs associated with trades. These transaction costs could make the cap and trade's costs higher than those of an equivalent carbon tax. However, most studies indicate that transaction costs are a small percentage of the value of allowances traded. According to the California Air Resources Board, in 2016, transaction costs under the linked California-Quebec

cap-and-trade program were about 2.2 percent of the value of traded allowances. Stavins (1994) offers an analysis of cap and trade in the presence of transaction costs. Studies suggest that transaction costs are very small compared with the value of the allowances traded. According to the California Air Resources Board, in the year 2016 transaction costs amounted to about 1.3 percent of the value of tradable allowances under the joint California/Quebec cap-and-trade program.

41. Under the carbon tax, when tax cuts are used to rebate revenues, the boost to the economy increases emissions, leading to a slight reduction in the emissions reductions relative to the policy with lump-sum rebates. Under a cap-and-trade program, when taxes are cut to refund auction revenue, the economic boost does not raise aggregate emissions—the aggregate cap is fixed in each year—but instead leads to higher demand for allowances and an increase in the market price of allowances. This higher market price slightly offsets the benefits of reducing preexisting distortionary taxes. The long-run price of allowances under lump-sum rebates is $60. The long-run price of allowances under payroll tax cuts, personal income tax cuts, and corporate income tax cuts are $60.75, $61.31, and $67.53, respectively.

42. Free allowances are not allocated to importers of refined products. They must still pay tariffs equivalent to the carbon content of their imports in the 100 percent free allocation scenario. This revenue is not enough to offset tax-base effects; thus, the net revenue from this policy is negative.

  Note that the policy simulation involved making up revenue by raising individual income taxes. The policy costs would be higher if instead the revenue loss were made up through an increase in corporate income tax rates, since corporate income taxes are more distortionary (have a higher marginal excess burden) than individual income taxes.

  It is possible to devise a policy involving 100 percent auctioning that involves the same cost as 100 percent free allocation. Suppose all the gross revenues from the auction are recycled lump sum to the private sector, and the needed revenue to make up the loss of revenue from other taxes is raised through an increase in individual income taxes. The costs of this (rather unlikely) policy would be very close to those under 100 percent free allocation.

43. These industries are oil extraction, natural gas extraction, coal-fired generation, other-fossil generation, petroleum refining, natural gas distribution, pipeline transportation, primary metals, and chemicals, plastics, and rubber.

44. During the 2017–2019 phase-in period, emissions allowance prices rise even faster.

45. The horizontal axis in the figure indicates percentage reductions in the present value of cumulative emissions over the infinite horizon.

46. When both the electricity-only and the economywide cap-and-trade program recycle the revenues via corporate income tax reductions, the economywide policy is more efficient at all levels of emissions reductions.

6. ALTERNATIVES TO EMISSIONS PRICING: CLEAN ENERGY STANDARDS AND A GASOLINE TAX

1. The RPS and CES are input-oriented intensity standards. Intensity standards can apply to outputs (or services performed) as well; these are often termed *performance standards*. Automobile fuel economy standards (floors on the miles per gallon) are an example.

2. See, for example, the analytical models in Holland, Hughes, and Knittel (2009) and Fullerton and Metcalf (2001). These models show that input-based intensity standards (of which the CES is an example) are formally equivalent to the combination of an emissions tax and input subsidy, with the implied revenue loss from the subsidy identical to the revenue gain from the tax. We make use of this equivalence in our model's representation of the CES.

3. For a detailed analysis of this issue, see Goulder, Hafstead, and Williams (2016).

4. As mentioned in chapter 4, a unit of output in the E3 model is the amount of a particular good that was worth $1 billion in 2013. Using data from the Energy Information Administration (EIA) on electric generation by source in megawatt hours, we calculate m by dividing 2013 megawatt hours by 2013 E3 output.

5. We include electricity from hydropower and nuclear power generators within the general classification of nonfossil electricity. In our analysis, all electricity generation within this classification qualifies as clean. As a result, our specification of the CES differs from the Bingaman proposal, which did not include hydro and nuclear within the clean category. Our analysis also differs from the Bingaman proposal in that we do not include the proposal's provision enabling utilities to pay a fine rather than meet the CES requirement.

6. The carbon tax is phased in over three years beginning in 2017 and is constant after 2019. We have also considered two alternative carbon tax profiles, one that matches *economywide* cumulative emissions reductions under the two types of policies, and one that matches *year-to-year* emissions reductions (electricity sector only or economywide). We have found that these alternatives yield very similar results for the relative costs of CES and carbon taxes.

7. The price index is defined in equation (3.13).

8. As with the cost curves displayed in chapter 5, the welfare measure is the negative of equivalent variation divided by the present value of tons reduced. This represents the average cost of reducing a ton of emissions.

9. The analytically solved general equilibrium model in Goulder, Hafstead, and Williams (2016) examines closely why the CES's costs rise relative to those of cap and trade as stringency increases.

10. We also find that carbon taxes are less costly when revenues are returned through cuts in individual capital taxes (a result not shown in the figure).

11. See chapter 5 for a discussion on how climate and nonclimate benefits are calculated.

12. The levels of emissions reductions are held constant in present value across policies. Because of differences in the timing of emissions reductions across policies, the calculated climate benefit will differ, reflecting the different social costs of carbon dioxide in different years.

13. This matches predictions offered by the analytically solved model in Goulder, Hafstead, and Williams (2016).

14. Changing the natural gas credit while holding fixed the required CES ratio can significantly change the stringency of the CES policy. In these simulations we compare CES policies that yield the same reductions in the present value of cumulative emissions as in our central case. This necessitated minor changes to the CES ratios, relative to the ratios used in the central case.

15. Source: Energy Information Administration, Federal and State Motor Fuels Taxes, February 2017.

16. See, for example, Leiby et al. (1997); Parry, Holmes, and Darmstadter (2003); and Brown and Huntington (2013).

17. It is worth noting that other policies could be introduced to achieve these nonclimate benefits. To reduce U.S. demands for crude oil imports, economists generally endorse an oil-import fee over a gasoline tax, since the fee would focus sharply on the item of interest: oil imports. Similar reasoning supports a congestion tax over a gasoline tax as an instrument for reducing road congestion. However, it is not clear that these policies would gain more political support than a gasoline tax increase. For some, an increase in the gasoline tax is a more feasible alternative.

18. According to the EPA, the combustion of 1 gallon of gasoline emits 0.0086 metric ton of carbon dioxide (1 gallon of diesel emits 0.01 metric ton). Therefore, a one-cent per gallon gasoline tax is equivalent to a $1.16 carbon tax on the carbon in gasoline. For diesel, the equivalent carbon tax is $1.00.

19. Under the 30-cent per gallon gasoline tax increase, the reductions in gasoline consumption relative to the reference case are about 6 percent. Under the 15-cent and 60-cent tax increases, the drops in gasoline consumption are approximately 3 and 11 percent, respectively.

20. As in the case under emissions pricing, windfall losses in the first period reduce capital gains tax revenues and reduce first period net revenue.

21. As with the carbon taxes previously considered, this carbon tax is set to phase in over three years, beginning in 2016. However, in contrast to previous cases, the tax rate remains constant after 2019.

22. For example, the 30-cent gasoline tax implicitly taxes the carbon content of gasoline at a rate of about $35 per ton; in contrast, in 2019 the rate of the equivalent carbon tax is much lower: $0.47 per ton.

23. The carbon tax delivers equal emissions reductions in present value but does not impose equivalent emissions reductions each period. As the per-ton value of climate benefits varies over time, differences in the timing of reductions lead to slightly different overall climate benefits.

24. See chapter 5 for a discussion of the energy security premium.

25. Emissions of NO_x and SO_2 contribute to ambient levels of $PM_{2.5}$. The monetized value of reduced NO_x and SO_2 accounts only for the benefits related to reduced concentrations of $PM_{2.5}$ and does not reflect any health impacts directly related to NO_x and SO_2

emissions or to other forms of pollution associated with NO_X (tropospheric ozone pollution) and SO_2 (acid rain).

26. The $364 billion estimate for the total benefit figure does not incorporate potential benefits from reduced traffic congestion.

7. DISTRIBUTION OF POLICY IMPACTS ACROSS INDUSTRIES AND HOUSEHOLDS

1. The effects on the economy as a whole differed significantly depending on the recycling method, but the form of recycling made little difference to the profit impacts in the carbon-intensive industries.

2. For further exploration using this framework, see Goulder, Hafstead, and Dworsky (2010). For analyses of some of these issues with somewhat different frameworks, see Parry (2003) and Burtraw and Palmer (2007).

3. In the future, carbon capture and sequestration might represent another channel for emissions abatement. We do not model this channel.

4. An analogy with imperfectly competitive markets yields further intuition for this result. Although restricting output can augment industry profits by causing an increase in the price of output, individual firms in a perfectly competitive environment gain nothing by reducing output: other firms would offset the reduction, and output prices would not rise. Imperfectly competitive firms, by contrast, can increase profits by *collectively* reducing output relative to the level under perfect competition. The imposition of the emissions cap causes firms to reduce output much as a cartel would, which works toward higher profits.

5. In decreasing order of the percentage loss of profit, these industries are coal-fired electricity generation, coal mining, natural gas extraction, other-fossil electricity generation, natural gas distribution, electricity transmission and distribution, pipeline transportation, petroleum refining, mining support, and railroad transportation. With the exception of railroad transportation, each industry has profit losses, in present value, in excess of 5 percent.

6. As in chapter 5, our cap-and-trade policy places a tariff on imported refined products. These revenues are collected under the 100 percent free allocation policy as well, but they are not enough to offset the tax-base effect; hence, the net revenues of this policy are negative.

7. The supply elasticities reflect the capital adjustment cost parameters in the model, while the demand elasticities reflect elasticities of substitution in production.

8. As noted previously, as a result of its negative impact on the tax base, the policy of 100 percent free allocation compels the government to raise these marginal rates. Hence for this policy, recycling through adjustments in individual income tax rates is more costly than lump-sum recycling.

9. Goulder, Hafstead, and Dworsky (2010) used a previous version of the E3 model in which the oil and natural gas extraction industries together formed a single "oil and gas" industry. As with the distinct oil extraction industry in the present version of E3, the earlier model accounted for stock effects (and associated implications for productivity)

in the aggregated oil and gas industry, and it treated foreign produced oil and gas as a perfect substitute for the domestic output of this industry. These assumptions led to smaller profit losses for the oil and gas industries under 100 percent auctioning than occur in the present model for these two industries together. Table 7.2 of this book indicates that about 23 percent of the allowances need to be given out free to preserve profits in the most vulnerable industries, considerably higher than the 13.7 percent reported in the Goulder, Hafstead and Dworsky study. The differences in results largely disappear if one disregards the 8.9 percent that must be devoted to the now-separate natural gas extraction industry.

10. Given the inframarginal nature of our allowance allocation scheme, firms' abatement decisions are largely independent of how many allowances they receive; this implies that the marginal cost of free allocation depends only on the total number of allowances allocated and not on the distribution of those allowances across industries. Hence a policy that allocates 10 percent of the allowance to one industry will have the same economy-wide cost as one that allocates 1 percent of the allowances to ten industries. However, the relative welfare costs of free allocation are decreasing in the stringency of the policy. In the central case, the cost of freely allocating 30 percent of allowances is 15 percent greater than the same policy with 100 percent auctioning, whereas the cost is 19 and 12 percent greater, respectively, in the low- and high-stringency cases.

11. The cap-and-trade policy in the fourth numbered column of Table 7.2 is not identical to the third numbered column of Table 7.3 because the level of the cap is set to equal the level of emissions under the carbon tax with lump-sum recycling. Under the Table 7.3 carbon tax with individual income tax recycling, the level of emissions is slightly higher than in the carbon tax case with lump-sum recycling. Thus, the carbon tax's cumulative emissions in column 3 of Table 7.3 are higher than under the cap in column 4 of Table 7.2. This difference accounts for the (very slight) differences in results.

12. For firms in these industries, the corporate income tax would still have incentive effects, since the corporate tax remains in place and thus the net receipt (value of the credit minus the gross corporate tax obligation) would depend on the firm's behavior.

13. Under this policy, the net revenue from the carbon tax exceeds the revenue lost from the directed rate cuts. To maintain revenue neutrality, the excess revenue is recycled through individual income tax cuts.

14. Providing corporate tax cuts to a subset of industries can have an adverse efficiency impact by introducing differences across industries in the tax rate, thereby creating an inter-industry distortion. In our simulations, any efficiency loss from such a distortion is not large enough to offset the beneficial impact from enabling some industries to enjoy a lower rate of corporate taxation, which in our model has a higher marginal excess burden than the individual income tax. Thus, providing compensation through targeted cuts in the corporate income tax proves to be very attractive on efficiency grounds.

15. The national value of shipments data yield different weights from those in the benchmark 2013 output data employed to calculate the aggregate profit impacts in table 7.3. In the absence of compensation (with individual income tax cuts), the average national profit

impact is −2.2 percent and −1.3 percent, respectively, with the value of shipments weights and the 2013 output data weights.

16. Other studies of the household distributional effects of carbon taxes include Dinan (2012); Mathur and Morris (2014); Rausch, Metcalf, and Reilly (2011); Fullerton, Heutel, and Metcalf (2012); and Williams et al. (2015). The first two articles in this list consider use-side impacts only; the latter three consider both the use and source sides.

17. Alternative specifications for household utility parameters, such as holding the elasticity of substitution for consumption and leisure constant across quintiles, do not substantially change our findings.

18. The benchmark labor endowment for each quintile is assumed to grow at the steady state growth rate, as in the E3 model, and is held fixed across reference and policy cases. For transfers and lump-sum taxes, we distribute the E3 level of transfer income and lump-sum taxes in each period to households using the shares used in the consistency procedure. Households are also endowed with their benchmark level of wealth in the first period. In subsequent periods, the level of wealth is endogenous and reflects the utility-maximizing saving behavior of each quintile.

19. Because the utility function is not consistent with perfect aggregation, the aggregate values of consumption, leisure, and wealth will not generally match values from the E3 model in the reference and policy cases despite the fact that our calibration approach ensures consistent aggregation. Yet we find the aggregation errors in the reference and policy cases to be extremely small.

20. Earlier studies also have tended to obtain regressive use-side impacts, although the earlier studies did not include attention to the influence of changes in the price of leisure. Nor did they consider how the impacts change over time.

21. In all present value calculations, we use the reference case nominal interest rate to discount future values back to the initial period of the policy.

22. Cronin, Fullerton, and Sexton (2017) found that carbon taxes with lump-sum rebates are generally progressive when (a) transfer income is inflation-indexed (as in the E3 model) and (b) annual consumption is used as a proxy for permanent income (as in the DH model when households are ranked by expenditure). Our analysis differs from this concurrent study by considering the household impacts under a broader range of revenue recycling options, including individual income tax cuts and corporate income tax cuts and allowing for general equilibrium impacts on factor prices and consumption quantities.

23. Under recycling that reduces the rates of preexisting taxes, the impact is progressive across quintiles 2 to 5, but the beneficial impacts is smaller for quintile 1 than for quintile 2. Because the first quintile has relatively little labor or capital income, it does not benefit much from the cuts to labor or capital taxes.

24. Under payroll tax cuts and individual income tax cuts, the impact is progressive across quintiles 2 to 5, but quintile 1 experiences a larger adverse impact than quintile 2. This reflects the smaller source-side impacts for quintile 1 relative to those for quintile 2.

25. As noted in chapter 6, the power-sector-only carbon tax offers a useful comparison since the CES applies to the power sector.

26. As in chapter 6, the equivalently scaled carbon tax profile is the one that generates the same cumulative reductions (discounted) in CO_2 over the infinite horizon.

27. The lowest quintile has the lowest spending share on motor vehicle fuels (4 percent), reflecting a lack of auto ownership for this group of households. Quintiles 3 and 4 display the highest fraction of spending on motor vehicle fuels (nearly 5 percent).

28. Cronin, Fullerton, and Sexton (2017) found that differences in carbon tax impacts within household quintiles might be as important as differences across quintiles. For example, they find that a majority of households in low-income deciles can have negative welfare impacts even when the average family in that decile has a positive welfare impact.

8. CONCLUSIONS

1. A recurring claim among climate scientists is that, to avoid extremely large climate-related damages, reductions in greenhouse gas emissions must be sufficient to prevent an increase of 1.5 degrees Celsius (2.7 degrees Fahrenheit) in global average surface temperature (Intergovernmental Panel on Climate Change 2014b). The Paris Climate Accord embraces this view, supporting the goal of "holding the increase in global average temperature to well below 2°C above preindustrial levels and pursuing efforts to limit the temperature increase to 1.5°C above preindustrial levels." The Intergovernmental Panel on Climate Change reports that to have a "likely chance" of limiting the increase in global mean temperature to 2°C, global greenhouse gas emissions would need to be cut by 40 to 70 percent from 2010 levels by mid-century, and to near zero by the end of this century (Intergovernmental Panel on Climate Change, 2014c). In our simulation experiments, we have considered cumulative reductions in U.S. emissions of up to 45 percent (from 2013 levels).

2. There is considerable uncertainty about the SCC. Earlier chapters considered a range of SCC time profiles.

3. We have not solved for the critical carbon tax profile under each recycling option, beyond which net benefits are negative. However, we would expect the critical stringency level to be lowest under lump-sum rebates and higher under the other, more efficient, forms of recycling.

4. Recall that gross costs are the economic costs before netting out the environment-related economic benefits.

5. The modest reductions also reflect the relatively low elasticity of demand for gasoline.

6. As illustrated by Figure 5.13 of chapter 5, the cost of achieving the central case (economywide) carbon tax's 29 percent reduction in cumulative emissions would be considerably higher under a power-sector-only carbon tax.

7. The gasoline tax policy we address covers only household purchases of motor vehicle fuels, excluding gasoline purchases by industry or commercial establishments. A broader policy would have more extensive profit impacts.

8. They are the most regressive because they produce the least progressive source-side impacts.

9. See, in particular, Fowlie, Reguant and Ryan's (2016) analysis of climate policy's impact on the imperfectly competitive Portland cement industry.

10. "Bottom-up" models are detailed engineering models that in most cases focus on one or two specific markets such as the market for electricity or natural gas. Some of these models capture significant details on production and transportation costs, and they often include market structures that differ from marginal cost pricing. The Haiku electricity sector model of Resources for the Future is an example of a bottom-up model.

11. See Borenstein and Davis (2012) for an analysis of the efficiency and equity implications of average-cost pricing in natural gas markets.

12. See, for example, Jaffe, Newell, and Stavins (1999), Gerlagh (2007); Gillingham, Newell, and Pizer (2008); Pizer and Popp (2008); and Acemoglu et al. (2012).

13. The particular policies that warrant removal remain a matter of debate. Some have suggested that the introduction of a nationwide CO_2 emissions pricing policy would justify eliminating the U.S. Corporate Average Fuel Economy (CAFE) standards. Others insist, however, that CAFE standards could still perform a useful service by addressing market failures that emissions pricing does not address. On this, see Davis (2016), Knittel (2011), and Stavins (2009). Similar debates arise over subsidies to low-carbon technologies.

APPENDIX A: FISCAL INTERACTIONS AND THE COSTS OF ENVIRONMENTAL POLICIES

1. For lump-sum rebates, the government revenues must be reduced by the value of the rebates. Therefore, government revenues $G^T = \tau_L(\bar{l} - \ell) + \tau_{x1}X_1 - \pi$ can be expressed by equation (A.8a) because carbon tax revenues equal the value of lump-sum rebates: $\pi = \tau_{X_1}X_1$.

2. Auctioning the allowances does not guarantee a revenue-recycling effect. This effect applies only when the auction revenues are recycled through cuts in the rates of a preexisting distortionary tax rather than via lump-sum rebates.

APPENDIX B: E3 MODEL STRUCTURE IN DETAIL

1. Industry subscripts are omitted for simplicity.

2. For a discussion, see, for example, Fagan (1997).

3. Stock effects also apply to natural gas production. We have not modeled stock effects in this industry based on the observation that technological advances in the production of shale gas are expected to offset significantly the negative productivity impacts of stock effects in this industry.

4. For all industries except housing, the tax rates τ^a and τ^h are equal. Only a small proportion of owner-occupied housing is taxable at the corporate interest rate, and τ^h reflects the average tax rate homeowners receive from mortgage interest deductions and local property tax deductions.

5. There is no uncertainty or risk in the model, and the arbitrage equation abstracts from risk. It is possible to modify the equation to account for risk differentials across assets.

6. Given the functional form of f and K, $f_{K^m,s} = f_{K,s} k_{m,s}$ where $f_{K,s}$ denotes the derivative of f with respect to total input K, and $k_{m,s}$ denotes the derivative of the capital aggregation function k with respect to capital stock K^m.

7. The function implies a unitary income elasticity, which is necessary for a balanced growth path in the presence of continued productivity growth.

8. By choosing next period's wealth W_{t+1}, the household is also implicitly choosing savings $W_{t+1} - W_t$.

9. The base of an ad valorem tax on a given consumer good is the value of spending on that good, while the base of an excise tax on a good is the quantity consumed of the good. Sales taxes are examples of the former; a gasoline tax is an example of the latter. Tax deductions are valued on the total level of consumption spending, inclusive of taxes, whereas credits are valued on the quantity of goods purchased.

10. Tax deductions reduce adjusted gross income. Hence $1 spent on a good that is subject to tax deductions lowers taxable income by the rate τ^s, where τ^s is a weighted average of labor income and capital income tax rates. In contrast with tax deductions, the value of a tax credit does not depend on the price of the good. We consider the prices of goods when calibrating the model's tax credits for the benchmark dataset, but the values of these credits are not affected by changes in policy.

11. If the constraint is not binding—that is, if $Z_t < A_t$, then the model's allowance price in period t is zero.

12. Motor vehicle fuels are not subject to ad valorem taxes or consumption subsidies.

13. We routinely check to verify that this condition is met when the other conditions are satisfied.

APPENDIX C: DATA, PARAMETERS, AND THE REFERENCE CASE PATH

1. As described in Chapter 4, data on the uses of foreign goods (intermediate inputs into production and final good demand) are used to create domestic and foreign IO matrices and use vectors. The aggregation and disaggregation methods apply to both sets of matrices and vectors.

2. For brevity, we refer to these vectors as "use vectors" because they are derived from the BEA Industry Use table.

3. These shares may seem arbitrary, but they were chosen to match data on fuel costs as a total of production costs and to assure that the value of the inputs approximates the value of the electricity produced by each generator.

4. Trade values are allocated to the "Trade" industry. Transportation values are allocated across transportation industries according to the share of each transportation industry's consumption expenditures, relative to total transportation consumption expenditures, in the BEA Use table.

5. We make no attempt to match prices year by year.

APPENDIX D: DISAGGREGATED HOUSEHOLD MODEL AND WELFARE CALCULATIONS

1. To solve the DH model we simultaneously solve for the transition of T periods and the steady state. We connect the two by linking period T wealth and the shadow value of consumption to the values of wealth and the shadow value of consumption in the steady state.

2. The full growth rate reflects both g_r, the steady-state real growth rate, and g_n, the steady-state growth rate of nominal prices: $(1+g_f)=(1+g_r) \cdot (1+g_p)$.

3. The reference case discount factor is used to avoid confounding differences in expenditures by period with changes in the interest rate.

REFERENCES

Acemoglu, Daron, Philippe Aghion, Leonardo Burstyn, and David Hémous. 2012. "The Environment and Directed Technological Change," *American Economic Review* 102(1), 131–66.

Aldy, Joseph E., and Robert N. Stavins. 2012. "The Promise and Problems of Pricing Carbon: Theory and Experience." *Journal of Environment and Development* 21(2): 152–80.

Anthoff, David, and Richard S. J. Tol. 2013. "The Uncertainty about the Social Cost of Carbon: A Decomposition Analysis Using FUND. *Climatic Change* 117: 515–30.

Armington, Paul S. 1969. "A Theory of Demand for Products Distinguished by Place of Production." *Staff Papers* 16(1): 159–78.

Bachmeier, Lance J., and James M. Griffin. 2006. "Testing for Market Integration of Crude Oil, Coal, and Natural Gas." *The Energy Journal* 27(2): 55–71.

Baker III, James A., Martin Feldstein, Ted Halstead, N. Gregory Mankiw, Henry M. Paulson Jr., George P. Shultz, Thomas Stephenson, and Rob Walton. 2017. "The Conservative Case for Carbon Dividends." Washington, D.C.: *Climate Leadership Council.*

Barrage, Lint. 2015. "Optimal Dynamic Carbon Taxes in a Climate-Economy Model with Distortionary Fiscal Policy." Working Paper, Brown University.

Bento, Antonio M., and Marc R. Jacobsen. 2007. "Ricardian Rents, Environmental Policy, and the 'Double Dividend' Hypothesis." *Journal of Environmental Economics and Management* 53(1).

Bento, Antonio M., Mark R. Jacobsen, and Antung Liu. 2015. "Environmental Policy in the Presence of an Informal Sector." Working Paper, Cornell University.

Borenstein, Severin, and Lucas W. Davis. 2012. "The Equity and Efficiency of Two-Part Tariffs in U.S. Natural Gas Markets." *Journal of Law and Economics* 55(1): 75–128.

Bovenberg, A. Lans, and Lawrence H. Goulder. 1996. "Optimal Environmental Taxation in the Presence of Other Taxes: General Equilibrium Analyses." *American Economic Review* 86(4): 985–1000.

——. 1997. "Costs of Environmentally Motivated Taxes in the Presence of Other Taxes: General Equilibrium Analyses." *National Tax Journal* 50(1): 59–87.

——. 2001. "Neutralizing the Adverse Industry Impacts of CO_2 Abatement Policies: What Does It Cost?" In *Behavioral and Distributional Effects of Environmental Policy*, eds. C. Carraro and G. Metcalf. Chicago: University of Chicago Press, 45–85.

Bovenberg, A. Lars, and Ruud de Mooij. 1994. "Environmental Taxation and Labor Market Distortions." *European Journal of Political Economy* 10(4): 655–83.

Bovenberg, A. Lars, and Frederick van der Ploeg. 1998. "Tax Reform, Structural Unemployment and the Environment." *Scandinavian Journal of Economics* 100(3): 593–610.

Boyce, James K., and Matthew Riddle. 2007. "Cap and Dividend: How to Curb Global Warming while Protecting the Incomes of American Families." Amherst: University of Massachusetts.

Brown, Stephen P. A., and Hillard G. Huntington. 2013. "Assessing the U.S. Oil Security Premium." *Energy Economics* 38: 118–27.

Bureau of Labor Statistics. 2014. "Consumer Expenditure Survey Public-Use Microdata." http:/www.bls.gov/cex/pumd_2013.htm.

Burtraw, Dallas, and Karen Palmer. 2007. "Compensation Rules for Climate Policy in the Electricity Sector." *Journal of Policy Analysis and Management* 27(4):819–847.

Carbone, Jared C., and V. Kerry Smith. 2013. "Valuing Nature in General Equilibrium." *Journal of Environmental Economics and Management* 66(1): 72–89.

Chen, Yunguang, and Marc A. C. Hafstead. 2016. "Using a Carbon Tax to Meet US International Climate Pledges." Resources for the Future Discussion Paper 16–48, Washington, D.C.

Chetty, Raj. 2009. "Is the Taxable Income Elasticity Sufficient to Calculate Deadweight Loss? The Implications of Evasion and Avoidance." *American Economic Journal: Economic Policy* 1(2): 31–52.

Congressional Budget Office. 2013. "The Distribution of Household Income and Federal Taxes, 2013." Washington, D.C.

Cooper, Russell W., and John C. Haltiwanger. 2006. "On the Nature of Capital Adjustment Costs." *Review of Economic Studies* 72(3): 611–33.

Coulson, N. Edward, and Herman Li. 2013. "Measuring the External Benefits of Homeownership." *Journal of Urban Economics* 77: 37–67.

Cremer, Helmuth, Firouz Gahvari, and Norbert Ladoux. 1998. "Externalities and Optimal Taxation." *Journal of Public Economics* 70(3): 343–64.

Cronin, Julie-Anne, Don Fullerton, and Steven Sexton. 2017. "Vertical and Horizontal Redistributions from a Carbon Tax and Rebate." Working Paper No. 23250: National Bureau of Economic Research.

Davis, Lucas. 2016. "New CAFE Standards: The Good, the Bad and the Ugly." Energy Institute at Haas Blog Post, energyathaas.wordpress.com, January 25, 2016.

Dinan, Terry. 2012. "Offsetting a Carbon Tax's Costs on Low-Income Households." Working Paper 2012–16, Congressional Budget Office.

Energy Information Administration. 2016. *Annual Energy Outlook*, DOE-EIA 0383. Washington, D.C.

Fagan, Marie N. 1997. "Resource Depletion and Technical Change: Effects on U.S. Crude Oil Finding Costs from 1977 to 1994." *The Energy Journal* 18(4): 91–105.

Feenstra, Robert C., Philip A. Luck, Maurice Obstfeld, and Katheryn N. Russ. 2014. "In Search of the Armington Elasticity." Cambridge, Mass.: NBER Working Paper 20063, April.

Feldstein, Martin. 1995. "The Effect of Marginal Tax Rates on Taxable Income: A Panel Study of the 1986 Tax Reform Act." *Journal of Political Economy* 103(3): 551–72.

Ferris, Michael C., and Todd S. Munson. 2000. "GAMS/PATH User Guide Version 4.3." Washington, D.C.: GAMS Development Corporation.

Fischer, Carolyn, Suzi Kerr, and Michael Toman. 1998. "Using Emissions Trading to Regulate US Greenhouse Gas Emissions: An Overview of Policy Design and Implementation Issues." *National Tax Journal* 51(3): 453–64.

Fowlie, Meredith, Mar Reguant, and Stephen Ryan. 2016. "Market-Based Environmental Regulation and the Evolution of Market Structure." *Journal of Political Economy* 124(1): 249–302.

Fullerton, Don, Garth Heutel, and Gilbert E. Metcalf. 2012. "Does the Indexing of Government Transfers Make Carbon Pricing Progressive?" *American Journal of Agricultural Economics* 94(2): 347–53.

Fullerton, Don, and Gilbert E. Metcalf. 2001. "Environmental Controls, Scarcity Rents, and Pre-Existing Distortions." *Journal of Public Economics* 80(2): 249–67.

Ganapati, Sharat, Joseph S. Shapiro, and W. Reed Walker. 2016. "Energy Prices, Pass-Through, and Incidence in US Manufacturing." Working Paper No. 22281: National Bureau of Economic Research.

Geisler, Charles and Currens, Ben. 2017. "Impediments to Inland Resettlement under Conditions of Accelerated Sea Level Rise," *Land Use Policy*, July.

Gerlagh, Reyer, 2007. "Measuring the Value of Induced Technical Change." *Energy Policy* 35: 5287–97.

Gillingham, Kenneth, Richard G. Newell, and William A. Pizer. 2008. "Modeling Endogenous Technological Change for Climate Policy Analysis." *Energy Economics* 30(6): 2734–53.

——. 1995. "Environmental Taxation and the 'Double Dividend:' A Reader's Guide." *International Tax and Public Finance* 2(2): 157–83.

Goulder, Lawrence H. 1998. "Environmental Policy Making in a Second-Best Setting." *Journal of Applied Economics* 1(2): 279–328.

Goulder, Lawrence H., and Marc A. C. Hafstead. 2013. "Tax Reform and Environmental Policy: Options for Recycling Revenue from a Tax on Carbon Dioxide." Resources for the Future. Discussion Paper 13-31. Washington, D.C: Resources for the Future.

Goulder, Lawrence H., Marc A. C. Hafstead, and Michael Dworsky. 2010. "Impacts of Alternative Emissions Allowance Allocation Methods under a Federal Cap-and-Trade Program." *Journal of Environmental Economics and Management* 60(3): 161–81.

Goulder, Lawrence H., Marc A. C. Hafstead, and Roberton C. Williams, III. 2016. "General Equilibrium Impacts of a Federal Clean Electricity Standard." *American Economic Journal: Economic Policy* 8(2): 186–218.

Goulder, Lawrence H., and Ian W. H. Parry. 2008. "Instrument Choice in Environmental Policy." *Review of Environmental Economics and Policy* 2(2).

Goulder, Lawrence H., Ian W. H. Parry, and Dallas Burtraw. 1997. "Revenue-Raising Versus Other Approaches to Environmental Protection: The Critical Significance of Preexisting Tax Distortions." *The RAND Journal of Economics* 28(4): 708–31.

Goulder, Lawrence H., and Roberton C. Williams III. 2012. "The Choice of Discount Rate for Climate Change Policy Evaluation." *Climate Change Economics*, November.

Hall, Robert. 1988. "Intertemporal Substitution in Consumption." *Journal of Political Economy* 96(2): 339–57.

Hall, Robert E. 2004. "Measuring Factor Adjustment Costs." *Quarterly Journal of Economics* 119(3): 899–927.

Hallegatte, Stephane, Mook Bangalore, Laura Bonzanigo, Marianne Fay, Tamaro Kane, Ulf Narloch, Julie Rozenberg, David Treguer, and Adrien Vogt-Schilb. 2016. *Shock Waves: Managing the Impacts of Climate Change on Poverty*. Climate Change and Development Series. Washington, DC: World Bank.

Heal, Geoffrey M., and Jisung Park. 2015. "Goldilocks Economies? Temperature Stress and the Direct Impacts of Climate Change." Cambridge, Mass.: National Bureau of Economics Research, Working Paper 21119.

Hoel, Michael, and Larry S. Karp. 2002. "Taxes Versus Quotas for a Stock Pollutant." *Resource and Energy Economics* 24: 367–84.

Holland, Stephen, Jonathan E. Hughes, and Christopher Knittel. 2009. "Greenhouse Gas Reductions under Low Carbon Fuel Standards?" *American Economic Journal: Economic Policy* 1(1): 106–46.

Hope, Chris. 2011. "The PAGE09 Integrated Assessment Model: A Technical Description." http://www.jbs.cam.ac.uk/research/working_papers/2011/wp1104.pdf.

Hughes, Jonathan E., Christopher R. Knittel, and Daniel Sperling. 2008. "Evidence of a Shift in the Short Run Price Elasticity of Gasoline Demand. *The Energy Journal* 29: 93–114.

Interagency Working Group on Social Cost of Carbon. 2013. "Technical Update of the Social Cost of Carbon for Regulatory Impact Analysis." In *Secondary Technical Update of the Social Cost of Carbon for Regulatory Impact Analysis*, ed. Secondary———. Reprint.

Intergovernmental Panel on Climate Change, 2014a. "Summary for Policymakers." In: *Climate Change 2014: Mitigation of Climate Change*. Cambridge, United Kingdom: Cambridge University Press.

Intergovernmental Panel on Climate Change, 2014b. *Climate Change 2014: Synthesis Report. Contribution of Working Groups I, II and III to the Fifth Assessment Report of the Intergovernmental Panel on Climate Change* [Core Writing Team, R.K. Pachauri and L.A. Meyer (eds.)]. IPCC, Geneva, Switzerland.

Intergovernmental Panel on Climate Change, 2014c. *Press Release: Greenhouse gas emissions accelerate despite reduction efforts* [Press Release]. Retrieved from: https://www .ipcc.ch/pdf/ar5/pr_wg3/20140413_pr_pc_wg3_en.pdf

Internal Revenue Service. 2015. Publication 1304, Statistics of Income Division. Washington, D.C., July.

Jaffe, Adam B, Richard G. Newell and Robert N. Stavins, 1999. "The Induced Innovation Hypothesis and Energy-Saving Technological Change," *Quarterly Journal of Economics* 114(3).

Joint Committee on Taxation. 2014. "Estimates of Federal Tax Expenditures for Fiscal Years 2014–2018." Washington, D.C.

Jorgenson, Dale W., Richard J. Goettle, Mun S. Ho, and Peter J. Wilcoxen. 2013. *Double Dividend: Environmental Taxes and Fiscal Reform in the United States.* Cambridge, Mass.: MIT Press.

Jorgenson, Dale W., and Peter J. Wilcoxen. 1996. "Reducing U.S. Carbon Emissions: An Econometric General Equilibrium Assessment." In *Reducing Global Carbon Dioxide Emissions: Costs and Policy Options*, eds. Darius Gaskins and John Weyant. Energy Modeling Forum, Stanford University, Stanford, Calif.

Kimball, Mike S., and Matthew D. Shapiro. 2010. "Labor Supply: Are the Income and Substitution Effects Both Large or Both Small?" Ann Arbor: Working Paper, University of Michigan.

Kling, Catherine, and Jonathan Rubin. 1997. "Bankable Permits for the Control of Environmental Pollution." *Journal of Public Economics* 64: 101–15.

Knittel, Christopher R. 2011. "Automobiles on Steroids: Product Attribute Trade-Offs and Technological Progress in the Automobile Sector," *American Economic Review* 2012, 101(December 2011): 3368–3399

Koskela, Erkki, and Ronnie Schöb. October 1999. "Alleviating Unemployment: The Case for Green Tax Reforms." *European Economic Review* 43(9): 1723–46.

Lawrance, Emily. 1991. "Poverty and the Rate of Time Preference: Evidence from Panel Data." *Journal of Political Economy* 99(1): 54–77.

Leiby, Paul N., Donald W. Jones, T. Randall Curlee, and Russell Lee. 1997. "Oil Imports: An Assessment of Benefits and Costs." Oak Ridge, Tenn.: Oak Ridge National Laboratory Discussion Paper.

Leiby, Paul, and Jonathan Rubin. 2013. "Energy Security Implications of a National Low Carbon Fuel Standard." *Energy Policy* 56, 29–40.

Li, Shanjun, Joshua Linn, and Erich Muehlegger. 2014. "Gasoline Taxes and Consumer Behavior." *American Economic Journal: Economic Policy* 6: 302–42.

Markandya, Anil, Mikel Gonzalez-Eguino, and Marta Escapa. 2013. "From Shadow to Green: Linking Environmental Fiscal Reforms and the Informal Economy." *Energy Economics* 40(Supplement 1): S108–S118.

Mas-Colell, Andreu, Michael Dennis Whinston, and Jerry R. Green. 1995. *Microeconomic Theory.* Vol. 1. New York: Oxford University Press.

Mathur, Aparna, and Adele C. Morris. 2014. "Distributional Effects of a Carbon Tax in Broader U.S. Fiscal Reform." *Energy Policy* 66: 326–34.

McClelland, Robert, and Shannon Mok. 2012. "A Review of Recent Research on Labor Supply Elasticities." Washington, D.C.: Congressonal Budget Office Working Paper 2012-12.

Metcalf, Gilbert E. 2017. "Implementing a Carbon Tax." Washington, D.C.: Resources for the Future.

Nordhaus, William D. 2007. "A Review of the Stern Review on the Economics of Climate Change." *Journal of Economic Literature* XLV.

——. 2014. "Estimates of the Social Cost of Carbon: Concepts and Results from the DICE-2013R Model and Alternative Approaches." *Journal of the Association of Environmental and Resource Economists* 1(1).

——. 2009. "The Economics of an Integrated World Market." International Energy Workshop Keynote Address, Venice, Italy.

——. 2014. "Estimates of the Social Cost of Carbon: Concepts and Results from the DICE-2013R Model and Alternative Approaches." *Journal of the Association of Environmental and Resource Economists* 1(1/2): 273–312.

Oates, Wallace E. 1993. "Pollution Charges as a Source of Public Revenues." In *Economic Progress and Environment Concerns*, ed. Herbert Giersch. Berlin: Springer-Verlag, pp. 135–52.

——. 1995. "Pollution Taxes and Revenue Recycling." *Journal of Environmental Economics and Management* 29(3): S64–S77.

——. 1997. "Environmental Taxes and Quotas in the Presence of Distorting Taxes in Factor Markets." *Resource and Energy Economics* 19: 203–20.

Parry, Ian W.H. 1998. "The Double Dividend: When You Get It and When You Don't." *National Tax Association Proceedings 1998*, pp. 46–51.

——. 2002. "Are Tradable Emissions Permits a Good Idea?" Resources for the Future, Issues Brief: 02-33.

——. 2003. "Fiscal Interactions and the Case for Carbon Taxes over Grandfathered Carbon Permits." *Oxford Review of Economic Policy* 19: 385–99.

Parry, Ian W. H., and Antonio M. Bento. 2000. "Tax Deductions, Environmental Policy, and the 'Double Dividend' Hypothesis." *Journal of Environmental Economics and Management* 39(1): 67–96.

Parry, Ian W. H., and Wallace E. Oates. 2000. "Policy Analysis in the Presence of Distorting Taxes." *Journal of Policy Analysis and Management* 19(4): 603–14.

Parry, Ian W. H., William Holmes, and Joel Darmstadter. 2003. "The Costs of US Oil Dependency." Washington, D.C.: Resources for the Future.

Parry, Ian W. H., and Kenneth A. Small. 2005. "Does Britain or the United States Have the Right Gasoline Tax?" *American Economic Review* 95(4): 1276–89.

Parry, Ian W. H., and Roberton C. Williams, III. 2010. "What Are the Costs of Meeting Distributional Objectives for Climate Policy?" *B. E. Journal of Economic Analysis & Policy* 10(2).

Parry, Ian W. H., Roberton C. Williams, III, and Lawrence H. Goulder. 1999. "When Can Carbon Abatement Policies Increase Welfare? The Fundamental Role of Distorted Factor Markets." *Journal of Environmental Economics and Management* 37(1): 52–84.

Pearce, David W. 1991. "The Role of Carbon Taxes in Adjusting to Global Warming." *Economic Journal* 101: 938–48.

Pizer, William A., and David Popp. 2008. "Endogenizing Technolgoical Change: Matching Empirical Evidence to Modeling Needs." *Energy Economics* 30(6): 2754–70.

Rausch, Sebastian, Gilbert E. Metcalf, and John M. Reilly. 2011. "Distributional Impacts of Carbon Pricing: A General Equilibrium Approach with Micro-Data for Households." *Energy Economics* 233: S20–S33.

Repetto, Robert, Roger C. Dower, and Jacqueline Geoghegan. 1992. *Green Fees: How a Tax Shift Can Work for the Environment and the Economy*. World Resources Institute, Washington, D.C.

Ross, Martin T. 2014. "Structure of the Dynamic Integrated Economy/Energy/Emissions Model." Working Paper, Nicholas Institute for Environmental Policy Solutions, Duke University.

Ross, Martin T., Allen A. Fawcett, and Christa S. Clapp. 2009. "U.S. Climate Mitigation Pathways Post-2012: Transition Scenarios in ADAGE." *Energy Economics* 31(Supplement 2): S212–S222.

Rubin, Jonathan, and Paul N. Leiby. 2013. "Tradable Credits System Design and Cost Savings for a National Low Carbon Fuel Standard for Road Transport." *Energy Policy* 56, 16–28.

Sedor, Noelle. 2015. "Why Fee and Dividend Is Better Than Cap and Trade at Fighting Climate Change." *Los Angeles Times*, March 5.

Smith, Anne E., Martin T. Ross, and W. David Montgomery. 2002. "Implications of Trading Implementation Design for Equity-Efficiency Tradeoffs in Carbon Permit Allocations." Working Paper. Washington, D.C.: Charles River Associates.

Smith, Anne E., and Martin T. Ross. 2002. "Allowance Allocation: Who Wins and Loses under a Carbon Dioxide Control Program?" Discussion Paper, Charles River Associates, Washington, D.C.

Smith, Anne E., David Harrison, Paul Bernstein, Scott J. Bloomberg, Will Gans, Noah Kaufman, Sebastian Mankowski, W. David Montgomery, Sugandha D. Tuladhar, and Mei Yuan. 2013. "Economic Outcomes of a U.S. Carbon Tax." Washington, D.C.: NERA Economic Consulting.

Stavins, Robert N. 1994. "Transaction Costs and Tradeable Permits." *Journal of Environmental Economics and Management* 29(2): 133–48.

——. 2009. "The New Auto Fuel-Efficiency Standards—Going Beyond the Headlines." Huffington Post Blog Post, www.huffingtonpost.com/robert-stavins/the-new-auto-fuel-efficie-b-206682.html, June 22, 2009.

Stern, Nicholas. 2008. "The Economics of Climate Change." *American Economic Review* 98.

Summers, Lawrence H. 1981. "Taxation and Corporate Investment: A Q-Theory Approach." *Brookings Papers on Economic Activity* 1: 67–127.

United States Bureau of the Census. 2015. "2012 Economic Census of the United States." Washington, D.C.

——. 2001. "Tax Normalizations, the Marginal Cost of Funds, and Optimal Environmental Taxes." *Economics Letters* 71: 137–42.

Weitzman, Martin L. 2010. "Risk-Adjusted Gamma Discounting," *Journal of Environmental Economics and Management* 60, 1–13.

Williams, III, Roberton C. 2002. "Environmental Tax Interactions When Pollution Affects Health or Productivity." *Journal of Environmental Economics and Management* 44(2): 261–70.

Williams, III, Roberton C., Hal G. Gordon, Dallas Burtraw, Jared C. Carbone, and Richard D. Morgenstern. 2015. "The Initial Incidence of a U.S. Carbon Tax across Income Groups." *National Tax Journal* 68(1): 195–214.

Wing, Ian Sue. 2006. "The Synthesis of Bottom-Up and Top-Down Approaches To Climate Policy Modeling: Electric Power Technologies and the Cost of Limiting US CO2 Emissions." *Energy Policy* 34(18): 3847–69.

INDEX

abatement, 160
  costs, 111, 134–35
  general equilibrium and marginal costs
    of, *21*, 21–22, 311*nn*11–12
  marginal, 82–83, 134, 139
adjustment costs, in E3 model, 41–42, 65,
  249, 250–51, 312*n*7, 315*n*15
ad valorem tax, 84, 328*n*12
  consumption and, 48, 66, *67*, 262–64,
  313*n*20, 328*n*9
  gasoline tax and, 167, 274
AEO. *See* Annual Energy Outlook
after-tax profits, 43, 252, 253, 273, 312*n*8
aggregate costs, of emissions pricing, 84
aggregation
  of BEA, 278–85, *280–82*, 328*n*1
  Leontief aggregation functions, 48,
  251, 262–63, 265–66, 294–95, 313*n*18
Air Quality Trends, 319*n*33
allowances
  alternative methods of allocation,
  187–88
  auctions of, 184, 186, 189, 219
  borrowing, 132–33
  in cap-and-trade system, 132, 141–42,
  182, 223, 319*nn*39–40, 320*nn*41–42
  distribution of, 129, 132
  distribution of profit impacts and,
  183–92

  emissions, 7, 14, 22, 51, 80, 125–26,
  319*n*39
  free allocation and, 184
  points of regulation and, 186
  prices and emissions, *134*, 135, *137*
  rents and profits and, 183–92, *184*. *See
  also* free allowances
alternative climate policies, distribution
  of profit impact under, 218–22,
  220–21
alternative parameter specifications,
  carbon tax, 121–23, *122*
alternative recycling methods
  carbon tax under, *111*, *117*, *220*
  GDP under, *220*
American Clean Energy and Security
  Act, 4
ammonia, 116
Annual Energy Outlook (AEO), 70,
  315*n*25
Annual Industry Accounts, 287
Armington trade elasticities, 48, 65–66,
  313*n*18, 315*n*19

base case time-path, 55, 69
BEA. *See* Bureau of Economic Analysis
behavior of firms, in E3 model, 42–44
benzene, 8
Bingaman, Jeff, 5, 144, 146, 321*n*5